I-MODE CRASH COURSE

Other McGraw-Hill Telecommunications Crash Courses

Bedell	*Wireless Crash Course*
Kikta, Fisher	*Wireless Internet Crash Course*
Louis	*M-Commerce Crash Course*
Shepard	*Optical Networking Crash Course*
Shepard	*Telecom Crash Course*
Shepard	*Telecommunications Convergence*

I-MODE CRASH COURSE

JOHN R. VACCA

McGraw-Hill
New York • Chicago • San Francisco • Lisbon
London • Madrid • Mexico City • Milan • New Delhi
San Juan • Seoul • Singapore • Sydney • Toronto

McGraw-Hill

A Division of The McGraw-Hill Companies

1 2 3 4 5 6 7 8 9 0 DOC/DOC 0 9 8 7 6 5 4 3 2 1

ISBN 0-07-138187-2

The sponsoring editor for this book was Steven S. Chapman, the editing supervisor was David E. Fogarty, and the production supervisor was Pamela Pelton. It was set in Fairfield by MacAllister Publishing Services, LLC.

Printed and bound by R. R. Donnelley & Sons Company

This book is printed on recycled, acid-free paper containing a minimum of 50 percent recycled de-inked fiber.

DEDICATION

To my little sister Madeline, who always triumphs and whom I love very much.

CONTENTS

FOREWORD

I-Mode and similar application platforms that will emerge during the next several years will play an important role in the advancement of wireless communications. It is important that product developers and managers move quickly to capitalize on the market potential that i-Mode offers.

I-Mode Wireless Crash Course provides an opportunity for everybody interested in this exciting new communications alternative to jump-start his or her understanding of i-Mode. The book covers i-Mode from the ground up, starting with fundamental concepts and terminology. The process of developing applications and where to find the best and most up-to-date information on this rapidly evolving technology are also provided. The technology side of i-Mode is rounded out with a thorough explanation of how to deploy i-Mode applications.

The product manager will greatly benefit on the chapters that cover the i-Mode market place. All readers will enjoy John Vacca's glimpse into the future of i-Mode.

There is much to do and the wireless market will not wait. I highly recommend that everybody who has an interest in i-Mode and wireless communications buy this book and get started today.

Michael Erbschloe
Vice President of Research
Computer Economics
Carlsbad, CA

ACKNOWLEDGMENTS

A very special thanks to my editor Steve Chapman, whose continued interest and support made this book possible. Thanks to my editorial assistant, Jessica Hornick, who provided staunch support and encouragement when it was most needed. Special thanks to my technical editor, Jim Helm, who ensured the technical accuracy of the book and whose expertise in mobile and telecommunications system technology were indispensable. Thanks to my Production Manager, Pamela Pelton; Editing Manager, David Fogarty; Project Manager, Beth Brown; all of the individuals at MacAllister Publishing Services, LLC; and, copyeditor, Jennifer Earhart, whose fine editorial work has been invaluable. Thanks also to my marketing manager, Elizabeth Schacht, whose efforts on this book have been greatly appreciated. And, a special thanks to Michael Erbschloe who wrote the foreword for this book.

Thanks to my wife, Bee Vacca, for her love, her help, and her understanding of my long work hours.

I wish to thank the organizations and individuals who granted me permission to use the research material and information necessary for the completion of this book. Finally, thanks to all the other people at McGraw-Hill Professional whose many talents and skills are essential to a finished book.

INTRODUCTION

Ever since its introduction in February 1999, i-Mode has been the most successful mobile internet technology in the world. By most successful, I don't mean the world-wide user number, but rather the quickness by which this technology has spread over Japan. If WAP would have boomed all over the world like i-Mode did in Japan, there would be approximately 800 million people who are using WAP on a regular basis today. In Japan, the number of i-Mode users is close to a sensational 35 million at the day of this writing. This means that 30 percent of Japan's total population is using i-Mode after not even three years of its existence!

The inventor of i-Mode is the Japanese company NTT DoCoMo. The start of i-Mode has turned many NTT DoCoMo stockholders into millionaires in an instant. The programming language for i-Mode is cHTML or i-Mode compatible HTML. *Compact HTML* (cHTML) is basically a trimmed down version of regular HTML. One can also make i-Mode sites with pure HTML if they know about the limitations of the i-Mode devices. For example, most i-Mode devices cannot display JPEG images or tables. The regular i-Mode page is very small and has only a few pictures to keep the download times as short as possible.

With that in mind, mobile internet technology is now fundamentally a part of whatever system you're building. In this book, in addition to i-Mode technology, I'll examine in detail many of these mobile internet systems, and show you how to build and maintain them.

PURPOSE

The purpose of this book is to show experienced (intermediate to advanced) mobile internet professionals how to quickly install i-Mode technology. It also shows through extensive hands-on examples, how you can gain the fundamental knowledge and skills you need to install, configure, and troubleshoot i-Mode technology. This book also provides the essential knowledge required to deploy and use i-Mode technology applications: integration of data, voice, and video. Fundamental i-Mode technology concepts are demonstrated through a series of examples where the selection and use of appropriate high-speed connectivity technologies are emphasized.

In addition, this book provides practical guidance on how to design and implement i-Mode applications. You will also learn how to troubleshoot, optimize, and manage a complex mobile internet using i-Mode technology.

In this book, you will learn the key operational concepts behind the mobile internet using i-Mode technology. You will also learn the key operational concepts behind the major i-Mode services. You will gain extensive hands-on experience designing and building resilient i-Mode applications, as well as the skills to troubleshoot and solve real-world mobile internet communications problems. You will also develop the skills needed to plan and design large-scale mobile internet communications systems.

Also, in this book, you will gain the knowledge of concepts and techniques that enable you to expand your existing mobile internet system, extend its reach geographically, and integrate global wireless network systems. This book provides the advanced knowledge that you'll need to design, configure, and troubleshoot effective i-Mode application development solutions for the Internet.

Through extensive hands-on examples (field and trial experiments), you will gain the knowledge and skills required to master the implementation of advanced residential i-Mode applications. In other words, in this book, you will gain the knowledge and skills necessary for you to take full advantage of how to deploy advanced residential i-Mode applications.

Finally, this intensive hands-on book provides an organized method for identifying and solving a wide range of problems that arise in today's i-Mode applications and mobile internet systems. You will gain real-world troubleshooting techniques and skills specific to solving hardware and software application problems in mobile internet environments.

SCOPE

Throughout the book, extensive hands-on examples will provide you with practical experience in installing, configuring,, and troubleshooting i-Mode applications and mobile internet systems. It will also provide you with advanced, extensive hands-on examples in configuring i-Mode applications and mobile internet systems. In addition to advanced i-Mode application technology considerations in commercial organizations and governments, the book addresses, but is not limited to, the following line items as part of installing i-Mode-based systems:

- The most obvious advantage of i-Mode is its color capability.
- I-Mode has been created out of some HTML-verifications which makes it a lot easier to handle for HTML-experienced programmers.
- The always-on technology makes fast access possible.
- Can be uploaded to a regular server.

- No special server, such as for WAP is needed.
- Can use most GIF-images, no need to change into a different format.

This book will leave little doubt that a new architecture in the area of advanced mobile internet installation is about to be constructed. No question, it will benefit organizations and governments, as well as their mobile internet professionals.

INTENDED AUDIENCE

This book is primarily targeted toward domestic and international network managers, technicians, designers and, consultants who are involved in designing, implementing, and troubleshooting i-Mode and other mobile internet systems. Basically, the book is targeted for all types of people and organizations around the world that are involved in planning and implementing i-Mode and other mobile internet systems.

PLAN OF THE BOOK

The book is organized into seven parts, as well as an extensive glossary of i-Mode and other mobile internet systems, 3G, and wireless internet networking terms and acronyms at the back. It provides a step-by-step approach to everything you need to know about i-Mode, as well as information about many topics relevant to the planning, design, and implementation of high-speed-performance mobile internet systems. The book gives an in-depth overview of the latest i-Mode technology and emerging global standards. It discusses what background work needs to be done, such as developing a mobile internet technology plan, and shows how to develop mobile internet plans for organizations and educational institutions. More importantly, this book shows how to install a mobile wireless broadband system, along with the techniques used to test the system and the certification of system performance. It covers many of the common pieces of mobile wireless broadband equipment used in the maintenance of the system, as well as the ongoing maintenance issues. The book concludes with a discussion about future i-Mode planning, standards development, and the wireless broadband mobile internet industry.

PART 1—OVERVIEW OF I-MODE WIRELESS TECHNOLOGY

Part 1 presents the fundamentals of i-Mode wireless technology: platforms; services and applications; marketing environment; and standards for next generation high-speed mobile wireless broadband connectivity.

Chapter 1: I-Mode Wireless Fundamentals This introductory chapter explores the widescale deployment, use, and services of i-Mode. It thoroughly discusses why the world's largest wireless Internet boom is happening in Japan, rather than North America or Europe.

Chapter 2: I-Mode Technology This chapter examines i-Mode technology (invented by Mari Matasunaga) as part of NTT DoCoMo's mobile Internet accessing system. It has become a great consumer success in Japan.

Chapter 3: Using I-Mode This chapter shows you how to use i-Mode technology. Today, DoCoMo's i-Mode cell phone service lets subscribers swap e-mail and pictures; search phone directories and restaurant guides; and download news, weather, and horoscopes.

Chapter 4: The Types and Sources of Technology That Support I-Mode This chapter shows you the types and sources of technology that support i-Mode. The future mobile phones will not be as different from current ones as had been thought. In other words, future generation mobile phones, for which operators have spent billions of dollars securing airwave space, will not prove the cash cows firms had hoped.

PART 2—DEVELOPING I-MODE APPLICATIONS

Part 2 of the book is the next logical step in i-Mode application development. Part 2 also examines i-Mode menu sites and how i-Mode is used for business.

Chapter 5: Official I-Mode Menu Sites This chapter discusses the official i-Mode menu sites. It examines how many i-Mode menu sites you can access, and what makes them so special. It all has to do with how the text on a i-Mode menu site is processed, so you can read it on a tiny cell-phone screen.

Chapter 6: Developing for I-Mode This chapter discusses developing for i-Mode. C HTML is recommended for small information appliances.

Chapter 7: I-Mode for Business This chapter discusses the use of i-Mode for business. Despite the success of the i-Mode mobile service for business, Internet use in Japan is abnormally low for a developed country.

PART 3—INSTALLING AND DEPLOYING I-MODE

Part 3 of the book discusses how to deploy i-Mode, implement i-Mode wireless broadband networks, and implement i-Mode wireless and satellite applications. It also examines i-Mode versus WAP technology; i-Mode emulator testing; i-Mode security; international i-Mode spectrum; e-mail, *short messaging system* (SMS), message free, and message-request; and 3G broadband mobile.

Chapter 8: I-Mode Versus Wireless Application Protocol (WAP) This chapter discusses i-Mode versus WAP—which is better? With wireless Internet usage in the United States falling short of analyst projections, many industry officials are eyeing the wildly successful i-Mode wireless Internet service rolled out by NTT DoCoMo Inc. in Japan.

Chapter 9: I-Mode Emulators: Testing This chapter discusses the testing of i-Mode emulators. There appears to be only one i-Mode emulator and it is in Japanese.

Chapter 10: Security on I-Mode This chapter discusses security on i-Mode. NTT DoCoMo Inc. has finally begun embedding digital certificates into its cell phones, with an eye to improving security for its 34.5 million i-Mode wireless Internet users.

Chapter 11: The International I-Mode Spectrum This chapter discusses the international i-Mode spectrum. Most of us are not enough of a sociologist to tell users if it's a human trait or just an American one, but in the States we always seem to be waiting for the knight in shining armor to save us, and that's also true in the wireless world.

Chapter 12: E-Mail, Short Messaging System (SMS), Message Free, and Message-Request This chapter discusses E-mail, *short message service* (SMS), message free, message-request. NTT DoCoMo, creator of the i-Mode wireless Internet service in Japan, has announced that it will begin offering its i-Mode service in the United States through AT&T Wireless in 2002.

Chapter 13: Third Generation (3G) Broadband Mobile This chapter discusses *third generation* (3G) broadband mobile communications. The proposed wireless systems may provide a unique solution for the highway portion of the high-bandwidth global communication system of the future.

PART 4—CONFIGURING I-MODE

Part 4 shows you how to configure i-Mode, JAVA, and i-Mode handsets.

Chapter 14: I-Mode and JAVA This chapter discusses i-Mode and JAVA. NTT DoCoMo recently entered into joint ventures to significantly expand i-Mode's presence in Europe and to support Java-based wireless services in Japan.

Chapter 15: I-Mode Handsets This chapter very briefly discusses i-Mode handsets. There is nothing really magic about it at all no proprietary handset technology, nothing that any other network operator couldn't replicate.

PART 5—MANAGING I-MODE MARKETS

Part 5 discusses how to manage i-Mode markets by taking a look at the overall i-Mode market, Japanese cellular market, and the Japanese mobile data market.

Chapter 16: Overview of I-Mode Markets This chapter presents an overview of i-Mode markets. Software companies see a wireless market potentially as valuable as that for personal-computer software.

Chapter 17: Overview of The Japanese Cellular Market This chapter presents an overview of the Japanese cellular market. Asia is filled with opportunity. However, the local players will become either your intimate business partners and your key to valuable markets, or they will become your fiercest competitors.

Chapter 18: The Japanese Mobile Data Market This chapter presents an overview of the Japanese mobile data market. When i-Mode makes the transition to 3G, some problems may be solved, such as the slow mobile data transmission speed.

PART 6—ADVANCED I-MODE FUTURE DIRECTIONS

Part 6 of the book discusses the future of i-Mode wireless broadband applications, markets, and handset suppliers. It also presents a summary, conclusions, and recommendations.

Chapter 19: The Future of I-Mode This chapter presents an overview of the future of i-Mode. While Americans look to cable modems and DSL to fulfill their broadband fantasies, Europeans want bandwidth for their mobile phones.

Chapter 20: Future Markets For I-Mode and Its Handset Suppliers This chapter discusses future i-Mode markets and its handset suppliers. Turning to the next-generation of off-the-wire communications, the lack of a specific air interface compels users to face choices in mobile services as they do in today's cellular market.

Chapter 21: Summary, Conclusions, and Recommendations This last chapter outlines the new challenges that mobile wireless brings to i-Mode service providers and addresses key properties that are critical to a robust mobile internet broadband provisioning system. Summary, conclusions, and recommendations with regards to the information presented in the book are also presented.

PART 7—APPENDICES

This last part includes eight appendices which contain various i-Mode resource listings, and a very extensive glossary of i-Mode, 3G, and mobile wireless internet terms and acronyms.

PART I

I-MODE WIRELESS FUNDAMENTALS

CHAPTER ONE

I-MODE WIRELESS FUNDAMENTALS

The teenage girls sit in cafes or postures on the corners of Tokyo's hip Shibuya district, dressed in short skirts and armed with fluorescent phones. When they flirt, they don't use their eyes. Instead, they deploy their thumbs, punching madly on the keypads of their mobile phones to dispatch an e-mail to a friend.

The hottest gadget in Japan, a phone that can ring up the Internet with the push of a button, is made by NTT DoCoMo[1], the hottest business in Japan. The company's name doesn't mean *anywhere* for anything. Its *information mode* (i-Mode) phones are ubiquitous in and around Tokyo, and the company makes no secret of the fact that it wants to conquer America next. Recently, it inked a deal with Internet giant America Online, essentially gobbling up AOL's struggling outpost in Japan and, more importantly, gaining a foothold in North America.

NOTE

NTT-DoCoMo is a subsidiary of Japan's incumbent telephone operator NTT. The majority of NTT-DoCoMo's shares are owned by NTT, and the majority of NTT's shares are owned by the Japanese government. NTT-DoCoMo's shares are separately listed on the Tokyo Stock Exchange and on the Osaka Stock Exchange, and NTT-DoCoMo's market value (capitalization) makes it one of the world's most valued companies.

However, will i-Mode translate into American? It is an open question whether the success of i-Mode is due to the superiority of the technology or is unique to behavior in Japan. No one disputes that cultural factors have helped fuel DoCoMo's success in Japan. Most Japanese do not own a PC, have never logged on to the World Wide Web and are completely clueless when it comes to streaming video, CD-quality music, and realistic online graphics. Having never taken a bite of the apple, DoCoMo's 26 million subscribers are content with the tiny, black-and-white screen on their mobile phones, which can do little more than display text and simple animation. Most of them don't even realize they are connected to the Internet. It's not the Internet to them; it's just a telephone with added features.

Internet-savvy Americans would never endure such dreary stuff, gloat naysayers. However, DoCoMo is testing international waters. It has a foothold in Indonesia and has purchased stakes in the Dutch company KPN Mobile and in Hutchison 3G UK Holdings, one of five licensees for next-generation phones in the United Kingdom.

What makes i-Mode a smash in Japan is its technical ingenuity and its got-to-have content (see sidebar, "Mad about i-Mode"). The design offers a huge variety of Internet sites and downloads, the phone displays them well enough, and the whole package is cheap. It uses a computer language that is a stepchild of the code used to design Web pages on the Internet. So, programmers accustomed to creating Web sites for the Internet can easily adapt them for access by i-Mode or even write new programs for it. The result is lots of content, including about 40,000 Web sites written just for the system, from horoscopes to train timetables to job-search engines.

NOTE

Mari Matsunaga invented i-Mode. Fortune Magazine *recently selected her as the most powerful woman in business.*

MAD ABOUT I-MODE

Understanding why some companies are agog over *mobile commerce* (m-commerce) takes no more than a look to Japan, where 16 percent of Japanese mobile phone owners use their phones to go online. Japanese wireless carrier NTT DoCoMo's i-Mode service has been a runaway success, so much so that the nation's largest telco stopped registering new users. Recently, Santa Clara, California-based enCommerce demonstrated that its getAccess software supported secure, wireless transactions on NTT DoCoMo, paving the way for m-commerce in North America (see Appendix E, "Internet, M-commerce, i-Mode In Japan").

Part of the reason for the success of the i-Mode service is that it addresses one aspect of Japanese culture-reticence. I-Mode is a fashion, but it's very suited as a way for Japanese people to communicate, because Japanese people don't like to say things directly. If you use i-Mode, you can say something more clearly than on the telephone.

Mobile messaging fits the Japanese culture and lifestyle much better than the desktop version—especially in the big cities like Tokyo. Everybody is always on the move; the trains are crowded, and the apartments and offices are small. Even small laptops are too big and unhandy to use on the train, but cell phones are small enough that you can hold them in one hand and even type with the same hand if necessary. You see people in Tokyo and throughout Japan typing mail messages on their phones without even looking at the keypad.

It's very easy to sign up and pay for i-Mode. Users can do it right on their phones with four clicks; the service is added to the monthly phone bill. Of course, unsubscribing is just as easy: you just delete the service from the special menu and that's it.

NTT DoCoMo's i-Mode is one of four wireless data services available in Japan. Though it's the most expensive, it is preferred because it's faster, has better service, and has more content. Although i-Mode may cost more than the others, it's still relatively cheap: 130 yen a week. That's analogous to U.S. wireless data services, which cost around $8 a month.

However, the analogy falls short when you compare the relative penetration of personal computers in the United States and Japan. They started with low penetration of the Internet in households and high penetration of wireless services. So when i-Mode came out, people could understand it easily even if they didn't have a home computer. The Unites States has the opposite—a higher penetration of household Internet usage.

continues

continued

However, a lesson is to be learn from Japan. DoCoMo is now the largest ISP in Japan because of its wireless Internet customers. Also, i-Mode was a success because it got content that customers wanted. That's thanks to the i-Mode's open architecture, which, like *Wireless Application Protocol* (WAP), is easy to build content for. Although that content will be different around the globe, you need content to be successful in wireless Net space.[2]

That's not the case in the United States and Europe, where DoCoMo's rivals, including Ericsson and Nokia, rely on a different language, called the wireless application protocol(WAP), that has yet to dazzle phone surfers (see Chapter 8, "I-Mode versus Wireless Application Protocol [WAP]"). The reason i-Mode is so popular is that it works well and it's cheap for the user.

Once online, subscribers pay for what they get, not for how long they're logged on. An e-mail message about the length of this paragraph can be sent for about two cents. DoCoMo also charges each time data is downloaded. That sets people back about 12 cents for a news item and 9 cents for a quick search of an English-Japanese dictionary.

Those yen do add up. The company earned $3.5 billion on $47.2 billion in sales in 2000. Yet, DoCoMo, in what may be its most important innovation, makes paying the bill painless. Signing up for the service is easy, and charges are simply added to the monthly phone bill, including the basic subscription charge, the cost of mail, and charges for visits to Web sites. That means you don't need a credit card every time you go online.

DoCoMo's technical wizardry doesn't mean that Internet phones will catch on in middle America, let alone catapult DoCoMo into an industry beater. As the offspring of *Nippon Telegraph and Telephone* (NTT), the near monopoly that dominates Japanese telecommunications, DoCoMo has built-in advantages at home that it won't get in the United States. NTT graciously set the table for DoCoMo: everything from handsets to the network infrastructure was designed to work in Japan (see sidebar, "Banking on Portable Phones").

BANKING ON PORTABLE PHONES

Recently, Japan's leading cellular-phone operator has signed up more than 4 million new subscribers for its mobile data service. People chattering away on mobile phones in restaurants and on street corners are hardly new. What is new in Japan is seeing people peering intently at mobile phone displays, laboriously punching the keypads, and then peering some more. People still use their cell phones for chatting, of course, but they are increasingly becoming tools for checking stock market news, transferring money, and ordering concert tickets.

Supporting most of this mobile financial activity is the new i-Mode service of *NTT Mobile Communications Network Inc.* (NTT DoCoMo). In addition to providing normal voice telephony, i-Mode phones can connect to an i-Mode server via a wireless packet data transmission network. The i-Mode server, in turn, connects to the Internet or directly to bank and brokerage servers. The phones have a simple text browser, slightly larger-than-usual displays and special function keys that enable users to move through text-based user interfaces.

I-Mode fills a need in Japan's communications market. *Personal computers* (PCs) have not been adopted by ordinary households to anywhere near the extent they have in the United States and Europe. In Japan, home PC users are still primarily hard-core computer hobbyists. On the other hand, four out of every five Japanese own portable phones, and they are ubiquitous among the more affluent young and middle-aged consumers most likely to be interested in financial services. A good match exists between cell phone users and the consumers service providers want to reach.

Launched in February 1999, i-Mode has attracted over 26 million subscribers as of this writing, with the number of new subscribers increasing by about 120,000 per week. So far, 92 banks are connected to the i-Mode network. Services vary slightly from bank to bank, but customers can generally open accounts, check account balances, transfer money between accounts, and pay bills.

continues

continued

Many of the banks offer similar services via PC connections through the Internet. Sakura Bank, for example, recently reported that over 71,800 customers made use of their i-Mode link. During the same time, only 61,100 customers used their PCs to access online services.

Although simply checking account balances generates the most traffic, surveys indicate that the service people value most is the interbank transfers that enable them to pay bills. Those subscribers using Sumitomo Bank's Ltd.'s i-Mode link make six such transfers a month on average.

The banking service is one of the most important i-Mode selling points, but in terms of daily hits, online banking ranks far below such i-Mode services as news, weather reports, and games. Most people don't check their account balances every day.

Many people do want to check their stocks every day. Six brokerages are currently offering regularly updated information on overall market trends and quotes on individual stocks. Customers with accounts can issue buy and sell orders via i-Mode.

For example, Daiwa Securities Ltd. in Tokyo currently has 9,000 customers per month using its i-Mode link. In addition to convenience, the i-Mode service is more secure than Internet-based access because of the dedicated, secure connection between the brokerage and the i-Mode server. Also, credit card companies use i-Mode to enable customers to check their account balances.

NTT DoCoMo plans even bigger things for i-Mode. More graphics and moving images will be possible with the increased bandwidth of next-generation mobile phone technology. Banks, brokerages, and credit card companies are also expected to be offering an ever-increasing menu of mobile financial services.

The United States is a different market. I-Mode, designed for NTT's network, can't simply be plugged into the AT&T or Sprint systems, which rely on different communications standards, different computer protocols, and even different phones. Also, though i-Mode's underlying language may be the darling of software programmers, the competing system, WAP, is more widely used and has a better chance of emerging as the basis of a world standard.

Judging from its moves elsewhere in the world, DoCoMo may begin with small steps in the United States. All they want to do is secure *roaming* agreements, so people can use DoCoMo handsets anywhere in the world. However, the company is also positioning itself to be a player in the United States when *third generation* (3G) wireless phones become reality. DoCoMo plans to be the first in Japan to introduce 3G mobile phones that will be able to receive data 40 times faster than today's i-Mode. The result will be high-quality sound, streaming video, and color graphics downloaded at lightning speed. The bet is that if its 3G phones are a hit, DoCoMo will have the clout to introduce its products in the United States.

However, will enough people use it? Besides playing e-mail tag, Japan's i-Mode fans love to download cute animated characters—cartoon versions of screensavers, really—and jingles to replace the standard ring of their phones. Those are hardly must-have services for American consumers. Not surprisingly, when IDC[3] asked mobile phone users if they cared to reach the Internet on their phones, three out of four said they were *very uninterested*. Wireless access to the Internet is inevitable and is a revolution. It just won't happen on a cell phone. DoCoMo is betting that's wrong.

What really is i-Mode? Let's take a closer look

WHAT IS I-MODE?

First introduced in Japan in February 1999 by NTT DoCoMo (as previously explained), i-Mode is one of the world's most successful services offering wireless Web browsing and e-mail from mobile phones. Whereas until recently, mobile phones were used mostly for making and receiving voice calls. I-Mode phones enable users also to use their handsets to access various information services and communicate via e-mail.

In Japan, i-Mode is most popular among young users, 24 to 35 years of age. The heaviest users of i-Mode are women in their late 20s. As of this writing, i-Mode had an estimated 25,028,000 user subscriptions.

When using i-Mode services, you do not pay for the time you are connected to a Web site or service, but are charged only according to the volume of data transmitted. That means that you can stay connected to a single Web site for hours without paying anything, as long as no data is transmitted.

How Did It All Start?

I-Mode is a mobile phone service that offers continuous Internet access. I-Mode is similar to WAP (WAP is another technology that has a scope of offering Internet access worldwide).

The reason DoCoMo decided to go with i-Mode instead of waiting for WAP is simple. The Japanese were ready to access the Internet through their mobile phones. They didn't want to have to wait for WAP to provide them with wireless data services they needed.

Consider that in 2000, NTT DoCoMo had 21 million subscribers and products that were preparing for the coming of *Wideband Code Division Multiple Access* (W-CDMA)—a technology that enables for the high-speed transmission of video and large-volume data. The Japanese (who make up the world's second-largest mobile phone market) were ready to access the Internet through their mobile phones. Thus, DoCoMo created i-Mode, along with a network of partners who offered specially formatted Web sites to fit into the small screen on the mobile handset.

NTT DoCoMo's decision to forego WAP for i-Mode was a completely practical solution. With 40 million new subscriptions predicted for the service by 2004, the decision was obviously the right one.

What's All the Fuss about?

Unless you've traveled to Japan, it is unlikely that you've directly experienced an i-Mode device. This part of the chapter explains what the fuss is all about.

I-Mode is foremost a brand, not a technology. As previously explained, this brand is owned by NTT DoCoMo, Japan's largest ISP. In some ways, i-Mode is equivalent to AOL: both are a brand representing a service or family of services. Until now, DoCoMo's advertising focused more on entertainment than on business applications. As is the case for the wired Web based mostly on PCs, the i-Mode "killer app" is e-mail, comprising nearly half of the total traffic.

If you look at the technology, the i-Mode service is based on packed switched overlay over circuit-switched digital communications. In contrast to most European or North American WAP services, it is based on TCP/IP, is always on, and hence does not require a dial-in connection. The content is encoded in an HTML variant named *Compact HTML* (cHTML).

From the marketing point of view, i-Mode is incredibly popular. It went from 0 to 11 million subscribers in about a year-and-a-half.

NOTE

About 80 million mobiles phone users are in Japan.

Mobile phones are less expensive than land phones, especially considering installation costs. Surprisingly, more WAP users are in Japan than anywhere else. In fact, the Japanese wireless Internet market is the biggest in the world. According to Eurotechnology, the approximate market share for wireless Internet is represented in Figure 1-1.[4]

Recently, DoCoMo entered into partnership with Sun to port Java. I-Mode devices will be able to run Java applets, enabling new kinds of applications like games or Web agents to

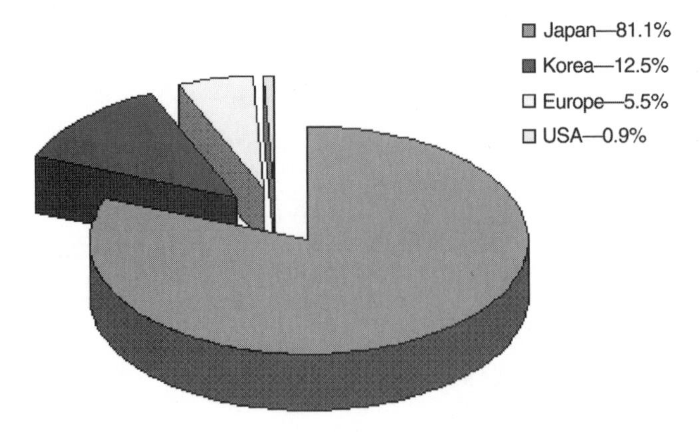

FIGURE 1-1 The wireless Internet.

be delivered to subscribers. At present, the i-Mode data transfer rate is 0.6 Kbps. In 2002, NTT DoCoMo will improve its infrastructure by moving to 3G wireless, with a data transfer rate of 384 Kbps for download and 64 Kbps for upload.

Again, in contrast to WAP devices that are still in black and white (or, green and black), some i-Mode devices have a color display with 256 colors. These devices can display *Graphic Interchange Formats* (GIFs) and even animated GIFs. Also, because only 5 to 7 percent of Japanese people speak English, i-Mode users are using Japanese. Most of the i-Mode handsets support phonetic text input. For instance, the number "1" is associated with the following phonemes: a, i, u, e, o. So it's probably easier to enter text with a phonetic base than it is with the alphabets; but seeing people typing the text, it doesn't seem to be any easier than with WAP devices. However, Japanese teenagers are typing on these devices like pros, and you should see how fast they are at writing e-mails by one-fingered typing.

Now, let's move on and briefly take a look at cHTML, the rendering technology in i-Mode.

cHTML: A Variant of HTML A cHTML document is like an HTML document. In contrast to a WAP document, which contains more than one screen (cards), a cHTML document contains only one screen. Thus, the cHTML-rendering model is identical to the HTML rendering model: one page at a time. An example of a cHTML document is as follows:

```
                    <html>
                    <head>
                    <META http-equiv="Content-Type"
content="text/html; charset=utf-8">
                    <META name="CHTML" content="yes">
                    <META name="description"
content="sample cHTML document">
                    <title>Sample cHTML document</title>
                    </head>
                    <body bgcolor="#ffffff" text="#000000">
                    <center>Didier's lab</center>
                    <hr>
                    Menu<br>
                    <A HREF="Messages" accesskey="1">1.</A>
Messages<BR>
                    <A HREF="Mail.htm" accesskey="2">2.</A>
Mail<BR>
                    <A HREF="http://www.eu-japan.com/i/"
accesskey="3">3.</A>
                    Eurotechnology cHTML page<BR>
                    <hr>
                    <center>
                    <A
href="martind@netfolder.com">email:martind@netfolder.com</a>
<br>
                    </center>
                    </body>
                    </html>
```

The first thing to notice is that cHTML is unfortunately not *eXtensible Markup Language* (XML)-based. This is why, in the preceding cHTML document, the element <hr> is not <hr/>. The cHTML language is similar to HTML and was submitted to the W3C as a Note in 1998. It was designed for low-memory footprint applications and so excludes things like tables and frames. The cHTML language has been adapted to the profiles of particular mobile devices by manufacturers. For instance, notice that the links are a bit different; a new attribute has been added: accesskey. This tells the phone browser to associate this link with a key on the keypad. It is easier to select a link with a single key press.

Because a cHTML document is so similar to an HTML document, an *Extensible Stylesheet Language Transformations* (XSLT) style sheet can be used to transform an XML document into cHTML in the same way that you already transform XML documents into HTML. Thus, the <xsl:output method="html"/> element can be used to specify the cHTML output type. For more information, see Chapter 6, "Developing for I-Mode," for a detailed discussion of cHTML.

So, what does an i-Mode-enabled phone look like? What does a typical i-Mode screen look like? Let's take a look.

TYPICAL I-MODE PHONE AND SCREEN

An i-Mode enabled cellular phone is similar in appearance to most cellular phone models. One feature in particular is a four-point command navigation button at the center of the phone. This enables the user to control the pointer on the display, as well as connect to the i-Mode service by pressing a single button.

Several companies manufacture i-Mode cellular phones, including Panasonic, Nokia (see example of phone as shown in Figure 1-2),[5] Ericsson, and Sony (these models are only available within the Asia Pacific). However, NTT DoCoMo's models are the most popular within the industry. Figure 1-3 shows the screen of Eurotechnology's site on a N502it (left) and a

FIGURE 1-2 Nokia's i-Mode-enabled phone.

imode screen (256 colors)

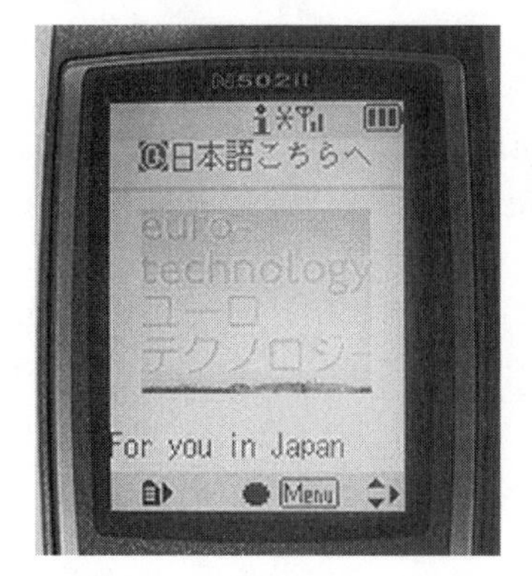

imode screen (black & white)

FIGURE 1-3 Typical i-Mode screens.

NM502i (right) handset characters (see sidebar, "Japanese Font and Proxy Service for I-Mode: Keitai-Font").[6]

NOTE

Around 90 or more different DoCoMo handsets (counting color variations) and many more for competing mobile Internet systems are in Japan.

I-MODE USERS

According to DoCoMo's data, the large majority of subscribers use e-mail and browse Web pages every day. Because 80 million mobile users are in Japan, i-Mode may potentially have around 80 million users in Japan. At the moment, 99.999 percent of i-Mode users are Japanese.

So, where are the world's wireless Internet users? Eighty-one percent (36 million) of the world's mobile Internet users are in Japan now, as shown in Figure 1-5![10] With the introduction of 3G broadband mobile services, Japan is now leading the world in mobile communications (see sidebar, "250 Times and Rising"). Can you afford to ignore Japan? I don't think so!

JAPANESE FONT AND PROXY SERVICE FOR I-MODE: KEITAI-FONT

Enfour Media Laboratory recently announced a new Japanese TrueType font with character set extensions specifically compatible with NTT DoCoMo's extremely successful i-Mode service. Accompanying this, a free proxy server service has been made available to enable display of nonstandard characters in all major desktop browsers for the first time. This makes for a unique combination destined to change your outlook on mobile content.

Called *Keitai-Font*, above the standard ShiftJIS 6,879 character set, the font includes 206 gaiji (nonstandard symbols and ideograph glyphs) and emoji (pictographs) included in Japanese mobile phones as shown in Figure 1-4.[7] A boon for cHTML content developers as well as casual *surfers*, Keitai-Font enables the viewing of i-Mode compatible content on Windows and Mac-OS desktops without the problem of missing character and garbage content. Keitai-Font fills the need made by the explosion in the popularity of e-mail and wireless Web content targeted to i-Mode users.

Note: Handsets using the i-Mode character set are used by 20 million people and growing daily.

Apart from the extended character set, Keitai-Font is unique in being a TrueType outline font based on 12 × 12 bitmap designs coming from the renowned digital design house TypeBank.[8] Making it useful even for just that different *bitmap look* in general, the *Data Transfer Process* (DTP) can be displayed and printed at any size. This makes it ideal for small screen banner-ad creation, printed manuals, screen mock-ups, handset prototyping, and other specialist imaging requirements. Keitai-Font supports outline creation and *Portable Document Format* (PDF) embedding. Keitai-Font is available on a hybrid CD costing 7,800 yen in Japan and $99 (U.S.) elsewhere including postage, handling, and insurance.[9]

NOTE

I-Mode users are a cross-section of Japan's society; i-Mode users include young people, but middle-aged and old people also use i-Mode.

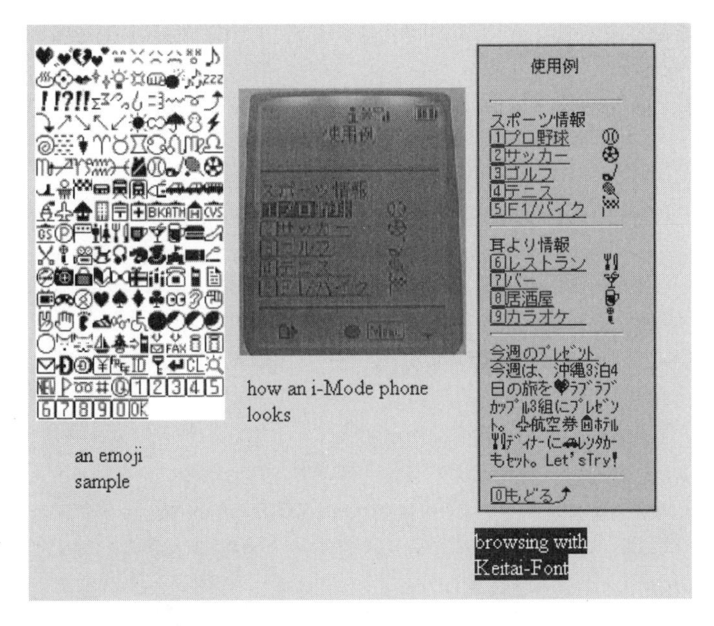

FIGURE 1-4 From left to right, the following shows: (a) an emoji sample, (b) how an i-Mode phone looks, (c) and how to browse with Keitai-Font.

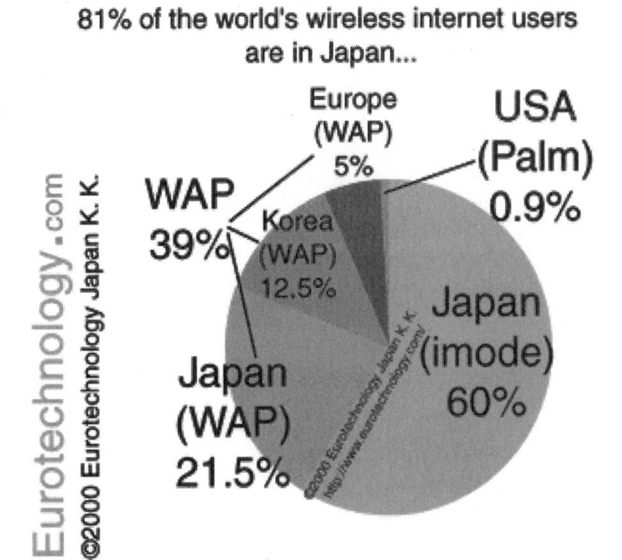

FIGURE 1-5 The world's wireless Internet users.

WHAT USERS DO ON I-MODE

Users send e-mail, look at the weather forecast, look at sports results, load ringing melodies into their handsets, play games, do online banking, online stock trading, purchase air tickets, download cartoons and images, look for restaurants, and look for new friends.

HOW DOES I-MODE WORK?

Technically, i-Mode is an overlay over NTT-DoCoMo's ordinary mobile voice system. Wheras the voice system is *circuit-switched* (you need to dial-up), i-Mode is *packet-switched*. This means that i-Mode is in principle *always on*, provided you are in an area where the i-Mode signal can reach you. When you select an i-Mode item on the handset *menu*, the data is usually immediately downloaded. There is no delay for dialing to set up the connection. However, there is a delay for the data to reach you. This delay is similar to the delay on your PC-based Internet connection after you click on a link or after you type in an URL and press the *return* or *enter* button. Of course, there are

250 TIMES AND RISING

The number of people accessing the Internet only by mobile phone totaled 7.63 million as of this writing, up more than 250 times from 30,000 in 2000. According to the report by Impress Corp., the number of people accessing the Internet stood at 43.74 million as of this writing, up sharply from 19.37 million in 2000.

The Internet-user population, which had increased by four to five million a year since 1997, increased explosively in 2000. Of the total Internet population, those accessing the Internet only by mobile phone accounted for nearly 20 percent, underscoring the popularity of wireless Internet access services, such as i-Mode offered by NTT DoCoMo Inc.[11]

further delays, if the information you download is too big, or if the network is overloaded.

How the I-Mode Network Service Works

The i-Mode service uses an additional packet communication network that is built onto DoCoMo's main network. This packet data transmission technology enables for constant connectivity. Thus, users are not charged for how long they are online, because this time is unlimited. Rather, users are charged only for how much information they retrieve.

With that said, essentially four main components are required for the i-Mode service. They are as follows:

- A cellular phone capable of voice and packet communication and with a browser installed
- A packet network
- An i-Mode server
- Information providers[12]

Unfortunately, the i-Mode service is currently only available in Japan and Hong Kong. However, plans are in the works to bring i-Mode to parts of Europe in the near future. It is unknown at the moment if i-Mode will make it to the United States.

How an I-Mode-Enabled Phone Connects to a Wireless Network

Typically, networks utilize two types of computers: servers and clients. *Servers* are the computers that hold information. *Clients* are the computers that you view the information from.

The way the Internet works is that servers hold your Web pages. You then view these Web pages from your PC (the client). In the case of the i-Mode Internet, an Internet server contains the i-Mode Web pages. Now, instead of viewing the pages from a PC, you are using a cellular i-Mode phone. These phones are now the clients.

Two other factors are involved in connecting a to a wireless network. In order to connect a cellular network to a server, a gateway must exist. Also, the Web site must be in an i-Mode format.

The Gateway A gateway translates wireless requests from a mobile phone to the server. It also sends information from a gateway back to the mobile phone. NTT DoCoMo provides a gateway to their users; however, this is only available to those in Japan.

Other gateways are on the market that enable users outside of Japan to build new mobile Internet services based on cHTML. One of the new gateways to hit the market is the m-WorldGate. This is the world's first commercially available cHTML gateway. m-WorldGate was developed by Logica.[13]

I-Mode Enabled Site

Web pages today are often written in HTML, which is too complex for mobile phones because of their slower connection speeds. An i-Mode enabled Web site utilizes pages that are written cHTML, which is a subset of HTML designed for devices with slower connection speeds.

Today, the i-Mode service boasts 600-plus i-Mode-enabled Web sites linked to a portal page, as well as 13,000 plus *unofficial* Web pages created by private individuals. What is cHTML like?

As previously explained, cHTML is extremely similar to HTML—in fact, it is HTML. The only difference is that some of the more resource-intensive areas of the code (such as tables and frames) have been taken out. Mobile devices have a slower connectivity speed. Thus, by eliminating some of the more involved portions of the code, cHTML enables i-Mode Web pages to download more quickly to mobile devices.

The World Wide Web Consortium (**http://www.w3.org**) contains a complete listing of the cHTML tags available to developers. Also, you can visit the NTT DoCoMo site at

http://www.nttdocomo/ser2.htm for an outline of cHTML tags available to developers.

WARNING

URLs are subject to change without notice.

WHAT I-MODE ENABLED WEB SITES LOOK LIKE Most i-Mode phones today utilize a micro-browser. These usually have a title bar with icons at the top of an LCD screen. These icons then enable users to access various services such as weather forecasts, transportation schedules, data searches, and news updates. Figure 1-6 is a text screen that displays text messages and data.[14]

One micro-browser in particular is Compact NetFront, developed by the Japanese company Access. Compact NetFront[15] is used as the micro-HTML browser for about 75 percent of all i-Mode-enabled devices. Information on Access's Compact NetFront can be found at **http://www.access.co.jp/english/product/proline/c_nf.htm**.

HOW TO DEVELOP AN I-MODE APPLICATION

The criteria for creating an i-Mode application or an i-Mode Web page are essentially the same as creating Web apps and Web pages with HTML. You must develop in the cHTML language, and then load the page to an Internet Web server utilizing FTP or some other transfer method.

FIGURE 1-6 Compact NetFront microbrowser.

Currently, the cHTML language does not support scripting language (this being a major obstacle for developers). However, NTT DoCoMo and Sun Microsystems have announced an alliance recently. There are plans to incorporate Java, Jini, and Java Card technologies into i-Mode cellular phones. The first wave of these phones is expected to hit the market around the end of 2002.

I-MODE EMULATORS

Yes, i-Mode emulators are available on the Internet. However, many handsets in circulation differ quite a lot. Some are color, others are black and white, and all are Japanese language. So at the moment, you need to be able to read Japanese in order to read most i-Mode content or to test i-Mode pages (see sidebar, "I-Mode Web-Based Emulator").

I-MODE WEB-BASED EMULATOR

The Wapprofit i-Mode emulator is a Web-based tool that acts like an i-Mode phone and can connect to a certain site to enable open access. Many i-Mode sites require a subscription, or they interrogate the device searching them and will display i-Mode content when they see a device code. It is currently free.

It is an easy to use products for Windows 95/98, Windows NT, Windows 2000, and XP systems. To use it live requires that you be connected to the Internet on a PC and have Microsoft Internet Explorer 5.0 installed.

The product consists of a simulated phone with a keypad that will work with approximately 90 percent of the sites that are coded i-Mode specification. You need to then enter the URL that is required in the top box. To get you started, you can enter **http://www.wapprofit.com/menu.chtml**.[16]

IMAGES AND MOVIES ON I-MODE

There are GIF images, animated GIF images, but no movies on i-Mode right now. Some handsets have color screens (256 colors) and can display color images.

ENCODING WEB PAGES FOR I-MODE

As previously explained, i-Mode uses cHTML, which is in part a subset of ordinary HTHL. However, in addition to HTML tags, there are some special i-Mode-only tags (for example, a tag to set up a link, which when pressed, dials up to a telephone number, or another i-Mode-only tag, informing search machines that a particular Web page is an i-Mode page).

In addition, there are also special DoCoMo characters, which are symbols for joy, kisses, love, sadness, hot spring baths, telephone, Shinkansen train, encircled numbers and so on. Quite a large number of these are special, nonstandardized characters.

LOOKING AT I-MODE WEB PAGES WITH YOUR ORDINARY NETSCAPE OR INTERNET-EXPLORER BROWSER

Of course, because cHTML is an extended subset of HTML, you can use your Netscape or *Internet Explorer* (IE) browser to look at i-Mode pages. Try it out and look at some of the i-Mode pages at the following URL: **http://www.eurotechnology.com/i/** or **http://www.eu-japan.com/i/**. However, at the moment, 99.999 percent of i-Mode users are Japanese and therefore almost all i-Mode content is in Japanese language (see sidebar, "I-Mode Content To Be Created by Disney"). Therefore, you will need a Japanese enabled browser; you will not be able to see i-Mode-only tags (such as the links that dial a telephone connection directly from the i-Mode handset in Japan), and you will not be able to see the many special DoCoMo i-Mode symbols. They will usually be replaced by a question mark. So, looking at an i-Mode page with an ordinary PC-based browser will give you an idea, but will not exactly reproduce what i-Mode users see on their handsets.

DIFFERENCE BETWEEN I-MODE AND WAP?

Comparing i-Mode and WAP is not straightforward. In some sense, i-Mode and WAP-based services are in competition in Japan and possibly world wide in the future. Both i-Mode and WAP are complex systems, and it is really only possible to compare present implementations of i-Mode and WAP, as well as their business models, the pricing, marketing and so on. Several

I-MODE CONTENT TO BE CREATED BY DISNEY

New Disney content has been delivered to Japanese i-Mode subscribers, and the company will deliver streaming video in 2002 after 3G service rolls out. The new *Disney-i* service includes *Character Town*, a daily Disney character screen saver, *Melody Palace*, a daily Disney song that plays on portable phones, *Game World*, the search for Disney characters in a virtual Disneyland, and *Disney Fan Magazine-i*, which provides Disney news and information.

Disney-i also launched additional entertainment options for i-Mode. In addition, the company promises streaming video when NTT DoCoMo rolls out faster 3G service in 2002.

This is an important business opportunity for Disney, particularly when you consider that the Japanese public has a strong association with Disney characters and with Tokyo Disneyland.

As you know, DoCoMo's i-Mode is wildly popular in Japan with more than 20 million subscribers. The Disney content is similar to that which already has proven popular among Japanese consumers.[17]

important differences exist in the way i-Mode and WAP-based services are presently implemented, marketed, and priced. As an example, i-Mode uses cHTHL, which is a subset of HTML, and is relatively easier to learn for Web site developers than WAP's markup language *wml*. Another difference is that at present in Japan, i-Mode is implemented with a packet switched system, which is in principle *always on* whereas WAP systems in Europe are at present circuit-switched (dial-up). Another difference is that at present, an i-Mode user is charged for the amount of information downloaded, plus various premium service charges (if used), while WAP services, are at present, charged by the connection time. Packet switching or circuit switching is a technical difference of the telecommunication system on which the services are based. It has nothing to do in principle with the i-Mode and WAP standards itself. In principle, i-Mode and WAP-encoded Web pages can be delivered over packet- and circuit-switched systems.

WHY I-MODE IS SO SUCCESSFUL

Not one single reason exists why i-Mode is so successful. It's to a large extent the fact that NTT-DoCoMo made it easy for developers to develop i-Mode Web sites, that PCs in Japanese homes are not so wide spread as in the United States and Europe, and that local access charges are very expensive in Japan, so that Japanese people don't use PCs for Internet access as much as in the United States or Europe, and several other reasons. The following is a list of possible reasons:

- Relatively low street price to consumers for i-Mode-enabled handsets at point of purchase.
- High mobile phone penetration (60 million mobile subscribers).
- Japanese people love gadgets.
- Relatively low PC penetration at home and high local loop access charges.

- i-Mode uses packet switched system: *always on* (if i-Mode signal reaches handset and so on), charges according to information accessed not usage time, relatively low fees.
- Efficient micro-billing system via the mobile phone bill. The micro-billing system makes it easy for subscribers to pay for value added, premium sites, and attractive for site owners to sell information to users.
- Fashion and efficient marketing.
- E-mail is the *killer ap*, like in the initial years of Internet growth.
- Uses cHTML, which makes it easy not only for developers but also for ordinary consumers to develop content. Explosive growth of content.
- AOL-type menu list of partner sites, and gives users access to a list of selected content on partner sites that are included in the microbilling system and can sell content and services.[18]

THE BANDWIDTH FOR DOWNLOADING DATA TO THE I-MODE HANDSET

The maximum speed for download is 9.6 Kbps. This is approximately six times slower than a 64-Kbps ISDN connection, but it is sufficient for simple i-Mode data. Of course, this speed makes it impossible to download live movies through i-Mode.

HOW TO RENT AN I-MODE PHONE

Can you rent an i-Mode handset for a week from a DoCoMo shop? No! You cannot normally, unless you have a residence permit and an *alien registration card* issued by Japan's Ministry of Legal Affairs (you need to be registered in Japan for a stay longer than three months normally). DoCoMo does not accept credit cards for bill payments.

However, if you just need a mobile telephone, it should be possible to obtain a prepaid mobile handset from one of the networks competing with DoCoMo. These, however, will not offer i-Mode access.

The Requirements

In other words, if you're coming to Japan (for a brief visit) and want to buy or rent an i-Mode phone, what are the requirements? The short answer is

- A Japanese visa that is valid for more than 90 days from the day of the phone purchase
- Your alien registration card
- Your passport[19]

On the other hand, the longer answer is as follows: Purchasing an i-Mode phone is easy provided you have an alien registration card and a visa valid for more than 90 days. That means that you can't get an NTT DoCoMo cell phone as long as you are just a tourist on a brief stay, even if you register yourself at your local ward office and have an alien registration card.

NTT DoCoMo will also not sell an i-Mode to you without registering you as a customer, which means you cannot just buy one as a souvenir. Basically, you need a working visa or a student/spouse visa in order to get an i-Mode phone.

To get a cdmaOne phone (WAP) is not that limited by the way. If you agree to pay with your credit card, you don't need to have the alien card and can sign up as a tourist as well with DDI and IDO.

To get prepaid phones is the easiest way, of course, (you don't have to sign any contract, and you pay in advance), but no i-Mode or WAP prepaid phones are on the market so far. The prepaid phones you can get offer just plain phone functionality: making and receiving calls.

CONCLUSION

This introductory chapter explored the wide-scale deployment, use, and services of i-Mode. It thoroughly discussed why the world's largest wireless Internet boom is happening in Japan, rather than North America or Europe. One factor in this may well be the billing mechanism. An important difference between the i-Mode billing system and conventional models is that with i-Mode, fees are based on the data transferred rather than connection time. Ten years ago, the Baby Bells did an experiment with Minitel-like terminals, with precisely the same billing system as the European and North American wireless network: time-based billing. From this experiment, they learned that North Americans prefer fixed payments. It seems that this lesson has been learned by wired Internet providers because most of them in North America charge a fixed fee. It seems that this lesson has not been learned by wireless Internet providers. The question to ask now is whether North American users prefer fixed charges for the wireless Internet? The answer to this question will have a large impact on the success of the wireless Web (see sidebar, "Could I-Mode Wireless Internet Be in the United States by 2002?").

Another issue to be resolved for WAP devices is screen size. It seems that most of the i-Mode devices available on the market offer bigger screen space than WAP devices. Surfing the wireless Web with this kind of device is more comfortable than with tiny three lines.

One thing is certain: although the wireless Web is taking shape in Japan and in Korea, Europe and the United States are lagging behind. Therefore, in order to take off in North America, the wireless Web needs to resolve these issues: the usage fees (fixed charges, minutes, per packets) and the screen size.

Perhaps the data input method needs more thought, too. However, after seeing Japanese teenagers adapting to one-fingered typing, maybe European and North American teenagers can adapt to it too. Who knows?

COULD I-MODE WIRELESS INTERNET BE IN THE UNITED STATES BY 2002?

With sales of mobile phones flagging and mobile-equipment makers slashing earnings estimates and laying off employees, NTT DoCoMo and AT&T Wireless are looking to jump-start wireless Internet access in the United States. The two companies plan to launch a U.S. version of Japan's popular i-Mode wireless Internet service early in 2002.

A subsidiary being created by the two telecom companies to support next-generation services in the United States hasn't even been formally created yet, and that no services could be rolled out until that subsidiary is in place. The source did say that some of the planned services will be based on NTT's i-Mode technology, which makes Web sites displayed on phones look more like a PC-based browser, but that the U.S. service may or may not feature the name i-Mode.

Analysts weren't surprised that the service is being planned, given that NTT in December 2000, acquired a 16 percent stake in AT&T Wireless for $9.8 billion. A key challenge will be getting consumers to upgrade their phones to i-Mode models. AT&T recently rolled out services tied to Nokia's new 8260 mobile handset, and the company isn't sure consumers will adopt the PC-upgrade mentality to their phone purchases simply to obtain better mobile Internet access. The jury is still out on how important wireless data is to the consumer market. However, the new service will let AT&T offer additional mobile applications to its business customers.

For an i-Mode-style service to succeed in the United States, American carriers will have to embrace the business model being employed in Japan, where consumers are willing to pay nominal subscription fees for news and data, and where NTT's billing systems let content providers bill consumers directly. I-Mode's popularity in Japan has been fueled in large part by the appetite for mobile content in the 15- to 25-year-old age group. That audience is more willing to pay for information, and U.S. wireless carriers must do a better job of addressing that demographic. So, if U.S. carriers pick up on the lessons of i-Mode, you should expect to see more enthusiasm for wireless Internet services here.

In any event, the next 20 chapters will thoroughly discuss in finite detail all of the topics examined in this chapter, and much, much more. Have a good read and enjoy!

END NOTES

[1]NTT DoCoMo, Inc., Sanno Park Tower from 27F to 44F, 11-1 Nagatacho-2-chome, Chiyoda-ku, Tokoyo, Japan, 2001.

[2]"In Japan, at Least, Wireless Rules," Business 2.0, 5 Thomas Mellon Circle, Suite 305, San Francisco, CA 94134, 2001.

[3]International Data Corporation (IDC), Five Speen Street, Framingham, MA 01701 USA, 2001.

[4]Eurotechnology Japan K. K., Parkwest Building 11th floor, 6-12-1 Nishi-Shinjuku, Shinjuku-ku, Tokyo 160-0023, Japan, 2001.

[5]"i-Mode: Introduction," AnywhereYouGo.com, 3000 Waterview Parkway, B2E14, Richardson, TX 75080, 2001.

[6]Eurotechnology Japan K. K., Parkwest Building 11th floor, 6-12-1 Nishi-Shinjuku, Shinjuku-ku, Tokyo 160-0023, Japan, 2001.

[7]"Keitai-Font," Enfour, Inc., 2-8-13 Sendagaya, Shibuya-ku, Tokyo 151-0051 JAPAN, 2001.

[8]TypeBank, Gyoen Plaza Bldg. 302, Daikyocho, Shinjuku-ku, Tokyo 160-0015, 2001.

[9]Enfour, Inc., 2-8-13 Sendagaya, Shibuya-ku, Tokyo 151-0051 JAPAN, 2001.

[10]Eurotechnology Japan K. K., Parkwest Building 11th floor, 6-12-1 Nishi-Shinjuku, Shinjuku-ku, Tokyo 160-0023, Japan, 2001.

[11]"Number of Wireless Internet Users In Japan Rises," AnywhereYouGo.com, 3000 Waterview Parkway, B2E14, Richardson, TX 75080, 2001.

[12]Ibid.

[13]Logica, Inc., 655 Third Avenue, Suite 1800-04, New York, NY 10017, 2001.

[14]AnywhereYouGo.com, 3000 Waterview Parkway, B2E14, Richardson, TX 75080, 2001.

[15]ACCESS Systems America, Inc., 860 Hillview Court, Suite 200, Milpitas, CA 95035, 2001.

[16]"Wapprofit: Mode Web-Based Emulator–Ver 1.1," Wapprofit, Technology Centre, Killarney.Co. Kerry. Ireland, 2001.

[17]INT Media Group, Inc., 23 Old Kings Highway South, Darien, Connecticut 06820, 2001.

[18]Eurotechnology Japan K. K., Parkwest Building 11th floor, 6-12-1 Nishi-Shinjuku, Shinjuku-ku, Tokyo 160-0023, Japan, 2001.

[19]WestCyber Corporation, 1350 Broadway, Rm. 705, New York, NY 10123, 2001.

CHAPTER TWO

I-MODE TECHNOLOGY

I-Mode consists of three technologies: a smart handset, a new transmission protocol, and a new markup language. This chapter explores the specifics of each technology.

SMART PHONE

A current high-end cell phone is now equivalent to a low-end PC. It has a 100-MHz processor, many megabytes of flash memory, and a color display with a graphical user interface. These *smart* phones enable users to browse the Net with a touch of button. However, users cannot talk while browsing the Web. They switch to the Web by hitting the URL with a button on the phone. No defacto standard is in the operation system and browsing software, such as Windows 2000 or the new Windows XP, or Internet Explorer. Because information that i-Mode deals with is still simple, each cell phone maker adopts its own system.

So what can you do with an i-Mode phone? For example, you can do the following:

- Reserve airline and concert tickets, find a good restaurant, check your bank balance or transfer money, read news and weather reports, check train schedules and city maps, download wallpaper images and ring tone melodies, and so on.

- Send and receive e-mail, not only to other i-mode users, but also to and from personal computers and handheld devices. When you subscribe to i-Mode, you automatically get an e-mail address that consists of your mobile phone number followed by @docomo.ne.jp.
- Access the Internet directly.[1]

TRANSMISSION SYSTEM

The transmission protocol of i-Mode is *Code Division Multiple Access* (CDMA), which enables several subscribers to use the same line at once. I-Mode's transmission speed is 9.6 Kbps, which is slower than a typical modem for personal use, 28.8 Kbps. Thus, e-mail is limited to about 250 characters per message. Although 9.6 Kbps is insufficient to download video, it is appropriate for short e-mail and simple graphics.

MARKUP LANGUAGE

I-Mode adopted *compact HTML* (C-HTML) as its markup language. C-HTML is a subtext of HTML, which focuses on text and simple graphics. Because a cell phone has a small display with touch button manipulation, it requires a special markup language to display data. Two major ways meet this need: C-HTML and *Wireless Application Protocol* (WAP). Web site operators can easily convert an existing Web page to C-HTML. As previously explained in Chapter 1, "I-Mode Wireless Fundamentals," WAP is an international standard language (major carriers that use WAP technology include: Ericsson, Motorola, Nokia, and so on). Because WAP is designed for a handset, users cannot read WAP pages from PCs.

INFORMATION PROVIDERS

DoCoMo helps its official content providers customize their i-Mode Web sites. Although subscribers can browse nonofficial

Web sites, official providers' sites can be accessed directly from DoCoMo's i-Mode menu, and official providers can change a monthly fee for a service. For example, the most popular official site is Bandai's download site for animations. Although the monthly fee is just one dollar, 1,100,000 subscribers pay $21.9 million to the company. Although i-Mode users can navigate among 6,000 cHTML sites, Japan has only about 300 WAP sites.

GENERAL TECHNOLOGY OVERVIEW

NTT DoCoMo seems to be planning to try to bring i-Mode to countries outside Japan. For example, DoCoMo, KPN, and Hutchison plan i-Mode mobile system in Europe. At the same time, other companies, such as China Unicom, are planning to offer a mobile Internet service similar to NTT DoCoMo's i-Mode.

I-Mode Speed

I-Mode phones transmit data at a speed of 9600 bps. Although this sounds slow compared to ordinary 56-Kbps computer modems, it is actually quite satisfactory for i-Mode, because e-mail is limited to only 500 bytes, and most i-Modes sites are relatively lightweight (made up mostly of text data with very few graphics, averaging about 1.2K in size). Downloading e-mail and i-Mode pages usually takes only a few seconds.

I-Mode Cost

Your bill will depend on how much you use i-Mode on your phone and whether you choose to sign up for any fee-based i-Mode content services. A basic fee costs 300 yen per month to access the i-Mode service. When you actually use i-Mode to surf Web sites and send or receive e-mail, you are charged 0.3 yen per packet (128 bytes) of transferred data (sent as well as received). According to NTT DoCoMo, the average total bill for i-Mode data transmission is about $13 (U.S.) per month.

Additionally, some i-Mode services charge a monthly fee, usually 100 to 300 yen per month. When you sign up for a fee-based service, you will be informed of the monthly fee before you are charged. Your monthly bill will combine the costs of your phone calls, packet data transmission fees, and any monthly fee-based services you've signed up for.

I-Mode-Compatible HTML

I-Mode-compatible HTML is based on a subset of HTML 2.0, HTML 3.2, HTML 4.0, and higher specifications that were extended by NTT DoCoMo with tags for special use on cell phones, such as the *tel:* tag, which is used to hyperlink a telephone number and let users initiate a call by clicking on a link (see sidebar, "Other HTML Tags or Attributes for I-Mode"). I-Mode-compatible HTML Web sites are easy to navigate because all basic operations can be performed using a combination of four buttons: Cursor forward, Cursor backward, Select, and Back (Return to previous page). Most phone browsers also provide a *Page Forward* button.

Functions that require two-dimensional navigation, such as image maps, and functions that require more intensive processing, such as frames and tables, are not included in the stan-

OTHER HTML TAGS OR ATTRIBUTES FOR I-MODE

Are there other special HTML tags or attributes for i-Mode? The following attributes were specially developed for i-Mode-compatible mobile sites:

<A href="tel:090******">call suzuki-san</A>**—This attribute enables users to select a telephone number link on their i-Mode phone and place a call to that number directly.

<A href="http://www.*.com/" accesskey='1'>Company Homepage </A>**—This attribute enables users to access a link by pressing a number on the keypad, rather than scrolling and then using the Select button, for faster navigation. In the preceding example, if the user presses the '1' button on his or her cell phone keypad, he or she will go directly to the company homepage.

PICTURE SYMBOLS

So, what picture symbols are allowed on i-Mode sites? The *istyle* attribute is as follows:

<INPUT type="text" name="aaa" value="@docomo.ne.jp" istyle =3 size=14 maxlength=20.

The *istyle* attribute sets the default character input mode (alphabetic, numeric, and so on.) as shown in Table 2-1.[2] The istyle attribute is not applicable to input type="password," whose input mode is always fixed as numeric. See Chapter 4, "The Types and Sources of Technology that Support I-Mode," for a further discussion in what picture symbols are allowed on i-Mode sites.[3]

TABLE 2-1 The *istyle* Attribute

ATTRIBUTE VALUE	MODE 1 (WHEN IN KANA INPUT MODE)	MODE 2 (WHEN IN PAGER MESSAGE INPUT MODE)
1 (default)	Full-space kana	Full-space characters
2	Half-space kana	Half-space characters
3	Alphabetic	Half-space characters (lower case recommended)
4	Numeric	Half-space characters (numeric recommended)

dard i-Mode HTML specifications. Here is an overview of some major features of HTML that are not included in i-Mode-compatible HTML:

- JPEG images
- Tables
- Image maps
- Multiple character fonts and styles
- Background colors and images
- Frames
- Style sheets[4]

NOTE

Some handsets do support features beyond the standard cHTML specifications, such as tables and multiple fonts. However, because these are not part of the published spec, you should use them with care and at your own risk!

Because i-Mode-compatible HTML is based on standard HTML, developers can make use of and adapt millions of HTML-based content resources, various software tools, and public materials (textbooks, magazines, and Web information). So, with that in mind, what kind of images can you use on i-Mode-compatible sites?

I-MODE-COMPATIBLE SITES IMAGES

Only GIF images are supported on i-Mode sites so far. For black-and-white screens, GIF images with the following characteristics are recommended:

- Black-and-white 2-bit GIF files
- Images in GIF 87, 87a, or 89a format
- Size: 94 × 72 dots (without scaling and scrolling on any screen)[5]

The dimension 94 × 72 is the size guaranteed to work on any handset. You can use GIFs larger than 94 × 72—some handsets have bigger screens (NM502i with 111 × 72 pixels) and can display bigger images. Smaller handsets will automatically reduce and dither images that are too large in an attempt to display them.

However, the total size of the page where the image is embedded (image plus text plus tags) must not exceed 5KB in file size. Color i-Mode displays can display GIF images based on an 8-bit (256-color) palette.

Using Animated GIFs

All i-Mode handsets from the second-generation 502i series, and some earlier models, can display animated GIFs with the following characteristics:

- Each animated GIF can consist of up to five frames.
- A maximum of four animated GIF files can be placed on the screen at one time.
- The maximum total size of image files that can be displayed on screen at any one time is 5KB.
- The maximum size for an animated GIF is 94 × 72 dots.
- An animation can be played up to 16 times. (When a number exceeding 16 is assigned, the animation plays 16 times and then stops.)
- The delay attribute makes it possible to set the interval between repeat plays in 1/100 increments. However, some 502i series models may be unable to reflect the interval adjustment with accuracy.
- The value of the delay attribute is based on a relative scale. (For example, delay=3 means the interval is three times longer than delay=1.)[6]

How Many Characters on how Many Lines Can Be Shown on an I-Mode Phone?

A minimum i-Mode display shows eight double-width characters (16 single-width, or English characters) on six lines without scrolling (see Figure 2-1).[7] Most i-Mode displays, however, are bigger than that (see Chapter 1). The largest screen to date (N502i with 118 × 128 pixel) can display 10 × 10 (20 ×10 single-width) characters without scrolling.

FIGURE 2-1 I-Mode display.

NOTE

English words do NOT wrap nicely (break at spaces) on i-Mode handsets—characters (Japanese and English) run to the end of a line, then continue on the next line, even if they break in the middle of a word. So, if you want nice word wrapping, you have to insert tags yourself, or else your text will look like Figure 2-2.[8]

Screen Size, Resolution, and Color of Different I-Mode Models

For a list of i-Mode handsets with their specifications (screen size, maximum number of full-width characters per screen, LCD dimensions, standby screen size, and display colors), please refer to the list published by NTT DoCoMo as shown in Table 2-2.[9] Here is an example:

- **I-Mode terminal** P501i
- **Screen size** 8 × 8
- **Number of full-width characters per screen** 64
- **LCD dimensions for browsing** 96 × 255 pixel
- **Standby screen size** 96 × 95 pixel
- **Display colors** Black and white[10]

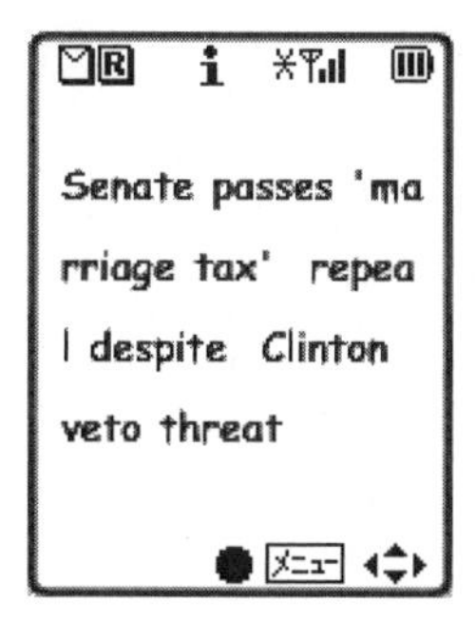

FIGURE 2-2 Basic i-Mode text.

I-Mode Phone Models that Are Available NTT DoCoMo is currently offering the following handsets with i-Mode support at: **http://www.mobilemediajapan.com/hardware/imode-handsets/.** An additional example is shown in Figure 2-3.[11]

Available I-Mode Accessories Mobile phone accessories like phone straps, flashing antennae, and mascots are extremely popular in Japan as shown in Figure 2-4.[12] You can buy a huge variety of phone-related accessories and items at almost any electronics shop in Tokyo. Some illustrated examples of cell phone accessories are shown here: **http://k-tai.impress.co.jp/column/strap/.**

Maximum Size Allowed for an I-Mode Page

The maximum size for an i-Mode page (including text and images) is 5KB; anything above that will not be displayed on the screen. DoCoMo, however, recommends a limit of 2KB per page.

Existing Constraints for URLs, Page Titles, and Bookmarks

What constraints exist for URLs, page titles, and bookmarks? The constraints are as follows:

- The maximum length of a character string is 200 bytes after URL encoding.

TABLE 2-2 Screen Size of I-Mode Terminals

I-Mode terminal	Screen size (max. no. of full-width characters per line × max. no. of lines)	Max. no. of full-width characters per screen	LCD dimensions for browsing	Standby screen size	Display colors
D501i	8 × 6	48	96 × 72 pixels		Black and white
F501i	8 × 6	48	112 × 84 dots		Black and white
N501i	10 × 10	100	118 × 128 dots		Black and white
P501i	8 × 8	64	96 × 120 pixels	96 × 95 pixels	Black and white
D502i	8 × 7	56	96 × 90 pixels	96 × 120 pixels	256 colors
F502i	8 × 7	56	96 × 91 dots	96 × 78 dots	256 colors
N502i	10 × 10	100	118 × 128 dots	(Large) 118 × 114 dots (Small) 118 × 70 dots	Four tone black and white
P502i	8 × 8	64	96 × 117 pixels	96 × 95 pixels	Four tone black and white
NM502i	8 × 6	48	111 × 106 pixels	95 × 76	Black and white
SO502i	8 × 8	64	120 × 120 dots	120 × 72 dots	Four tone black and white
F502it	8 × 7	56	96 × 91 dots	96 × 78 dots	256 colors
N502it	10 × 10	100	118 × 128 dots	(Large)	256 colors

				118 × 114 dots (Small) 118 × 70 dots	
SH821i	8 × 6	48	96 × 78 dots	96 × 78 dots	256 colors
N821i	10 × 10	100	118 × 128 dots	(Large) 118 × 114 dots (Small) 118 × 70 dots	Four tone black and white
P821i	10 × 10	100	118 × 128 dots	(Large) 118 × 114 dots (Small) 118 × 70 dots	Four tone black and white
D209i	8 × 7	56	96 × 90 pixels	96 × 120 pixels	256 colors
F209i	8 × 7	56	96 × 91 dots	96 × 78 dots	256 colors
N209i	9 × 6	54	108 × 82 dots	(When the time is not displayed) 108 × 82 dots (When the time is displayed) 108 × 54 dots	Four tone black and white
P209i	8 × 6	48	96 × 87 pixels	96 × 91 pixels	Four tone black and white
P209iS	8 × 6	48	96 × 87 pixels	96 × 91 pixels	256 colors

*The LCD size includes the display area for images.

FIGURE 2-3 Handset with i-Mode support.

FIGURE 2-4 Cell phone accessories.

- The maximum length of an URL that can be input directly is 100 bytes.
- The maximum length of an URL that can be added to the bookmark list is 100 bytes.
- The maximum length of the title of a page/bookmark is 24 bytes.[13]

User Agent of I-Mode Models

Finally, what does the user agent of i-Mode models look like? All i-Mode HTTP_USER_AGENT strings have the format

```
DoCoMo/HTTP version/model name
Example: DoCoMo/1.0/F501i.
```

The new 502 generation additionally transfers some information about the cache if the cache size is increased:

```
DoCoMo/HTTP version/model name/cache
```

For example, DoCoMo/1.0/F502i/c10, means that an F502i terminal is accessing the site with its cache size increased to 10KB from the standard 5KB.

CONCLUSION

This chapter examined i-Mode technology (invented by Mari Matasunaga) as part of NTT DoCoMo's mobile Internet accessing system. It has become a great consumer success in Japan. This system, which enables users to access everything from e-mail to restaurant guides to interactive games on their mobile phones, started with zero users in February 1999. It now has as many as 40 million subscribers and the potential to be a leader in Internet service.

Technically, i-Mode is an overlay of NTT DoCoMo's ordinary mobile voice system. Although the voice system is circuit-switched (that is, you need to dial-up), i-Mode is packet-switched. This means that i-Mode is always on, in principle, provided you are in an area where the i-Mode signal can reach you. Users send e-mail, look at weather forecasts, view sports results, load music into their handsets, play games, do online banking and online stock trading, and look for restaurants and new friends. The i-Mode technology uses cHTML, a subset of HTML, and is easier for Web site developers to learn than the *wireless markup language* (WML). I-Mode Java™-enabled telephones can use the currently deployed i-Mode network architecture. Downloads from sites will be handled through cHTML, and the HTTP protocol, uploads, and data transfers from i-Mode Java-enabled telephones also will be handled by cHTML.

END NOTES

[1]WestCyber Corporation, 1350 Broadway, Rm. 705, New York, NY 10123, 2001.

[2]Ibid.

[3]Ibid.

[4]Ibid.

[5]Ibid.

[6]Ibid.

[7]Ibid.

[8]Ibid.

[9]NTT DoCoMo, Inc., Sanno Park Tower from 27F to 44F, 11-1 Nagatacho-2-chome, Chiyoda-ku, Tokoyo, Japan, 2001.

[10]Ibid.

[11]WestCyber Corporation, 1350 Broadway, Rm. 705, New York, NY 10123, 2001.

[12]Ibid.

[13]Ibid.

CHAPTER THREE

USING I-MODE

I-Mode is ubiquitous in Japan, where millions use the service to enable their mobile phones to exchange e-mail messages, download images such as screen savers, play games, retrieve stock prices, find restaurants, and find movie information. NTT DoCoMo, which owns a 16-percent stake in AT&T Wireless, expects to roll out i-Mode in the United States at the beginning of 2002, starting in Seattle. [1]

However, analysts are wary that i-Mode's success in Japan may not repeat itself in the United States because of sociological differences between the cultures. One of the distinguishing factors in the Unites States is the heavy use of private transportation versus public transportation. A significant amount of i-Mode usage takes place while people are commuting, but in the United States, we spend more time in private transportation than public transportation.

NTT DoCoMo also has the advantage of a 60-percent market share in Japan, whereas in the United States, more carriers are competing for subscribers. Perhaps the biggest reason i-Mode won't duplicate its success is that the Japanese are much more dependent on mobile services for Internet access. Only 20 percent of the Japanese access the Web on a desktop computer, so most opt to access it from their phones. However, Americans have access to cheaper computers and opt for machines with larger screens and keyboards.

Americans are also not used to paying for Internet content, whereas NTT DoCoMo has had no trouble charging $1 to $3 a month for content such as ring tones and screen savers. NTT

DoCoMo receives nine cents for every dollar paid for content, and the content provider takes the rest.

NTT DoCoMo currently touts 20 million i-Mode subscribers. If you provide good content, you can make money like the regular Internet.

Analysts indicate i-Mode could be decently adopted in the United States only if NTT DoCoMo continues its marketing strategy of not branding its service as the *wireless Web* or a *mobile Internet*, which could mislead Americans into believing they're getting the services of a miniature desktop computer.

NTT DoCoMo never sold i-Mode as the wireless Web. I-Mode is, "Do you want to buy stuff? Do you want the weather and stock quotes?" They never said,"Do you want to surf the Web?"

NTT DoCoMo also never confused consumers with different protocols and different ways to deliver the data on the phones. On less bandwidth than slower dial-up modems in America (9.6 Kbps), it was able to deliver an affordable robust service.

Here you have a carrier on its own that indicates you are going to build a wireless business on your own type of browser with low bandwidth. You are also going to put forward a compelling content package and market it, so people would want to use it. NTT DoCoMo didn't get caught in if WAP is better than i-Mode or if XML is better than that. They built a business.

The amount of content written in i-Mode's language, *compact HTML* (cHTML), is what has made it stick with consumers. People will pay for the content if it's worth paying for it. If you could pay $1 a month to see if your flight will be delayed, you'd pay for it.

Americans will access the exact services available in Japan. Currently, 40,000 Japanese-language i-Mode sites are written in cHTML or a stripped-down version of HTML.

You have to be very careful. The Americans and Japanese are different.

However, the service rates would remain the same—an average of $70 a month for voice services and $17 for i-Mode transmission fees.

The i-Mode-enabled phones will run on a packet-switched network as in Japan, which means the phones are always on and don't have to be recharged for at least a week. When subscribers access content on the phones, they pay for the amount of data they download rather than for airtime.

The phones that run on the service are sexy—most of the phones are sleek and silver, weigh almost two ounces, have full-color screens, and support Java applications. Nevertheless, many researchers in the industry don't know for sure whether i-Mode will tout the same support as in Japan. However, it could give other wireless Web services, such as WAP, a run for their money. I-Mode's entrance would settle that barroom argument.

So, with all of the proceeding in mind, how do you use i-Mode, how do you look at i-Mode pages, what can you find on i-Mode, what can you do with i-Mode, and so on . . . ? Let's take a look.

WHERE I-MODE CAN BE USED

At the moment, i-Mode can only be used in Japan. Within Japan, you can use i-Mode in the most highly populated areas, along major motorways, and in some subway stations.

OUTSIDE OF JAPAN

Up until now, i-Mode service is provided only on NTT DoCoMo's Japanese network. You can, of course, publish i-Mode-compatible Web sites outside Japan on any Web server in the world. To see and use i-Mode services on a cell phone, however, you will have to be in Japan.

SUBSCRIBING TO I-MODE

You need to obtain an i-Mode-enabled NTT-DoCoMo handset directly at one of the many DoCoMo shops or at shops that sell DoCoMo handsets and contracts. You also need to enter into a service agreement for DoCoMo's cellular phone service, and you need to choose the additional i-Mode option.

HOW I-MODE WORKS AND HOW YOU CAN GET STARTED

Your DoCoMo handset will have a special i-Mode button. You press this i-Mode button, and the i-Mode menu appears. On the i-Mode menu, you can choose one of several options: you can look at your private *my menu* page, where i-Mode sites are listed that you are subscribing to, or you can look at the menu of approximately 600 i-Mode partner sites, or you can type in an URL to look at any Internet site, bookmark Web sites, or you can send e-mail or change your settings.

COST OF AN I-MODE PHONE IN JAPAN

Consumers in Japan don't pay the true cost of i-Mode-enabled phones. Typical total costs (including DoCoMo's charges) when switching to i-Mode are on the order of $70, but this can be lower (even down to zero in extreme cases) or higher, depending on the popularity of the model or the geographic area of Japan.

Cost To Use I-Mode

To use i-Mode, you have to have a basic mobile phone subscription with NTT-DoCoMo. On top of the basic subscription costs, you have to pay 300 yen (approximately $3) per month for i-Mode use. This is all you pay if you never actually use i-Mode. However, when you start using i-Mode, you incur additional charges. There is a basic data charge per packet, 0.3 yen (approximately $0.3) per data packet transmitted at 128 bytes. As an example, looking at the basic i-Mode menu, the standard DoCoMo welcome screen, or user interface, will set you back about 2.7 yen (approximately $2.7). I-Mode doesn't charge for connection time. In addition, there are other charges for using e-mail and for premium subscription services.

LOOKING AT INTERNET SITES THAT ARE NOT SPECIFICALLY FORMATTED FOR I-MODE

In principle, you can look at Internet sites that are not specifically formatted for i-Mode. However, in many cases, you will only see a small portion of the page, and the information you see may not be meaningful. Also, it is likely that the information will overflow the cache of your handset; in this case, the displayed information will be truncated, and there will be an error message as well.

PREMIUM PAY SITES

Many (but by far not all) of DoCoMo's *official* partner sites are pay sites. These sites will have a public free area for basic information, but most of the content will require that you register and pay a monthly charge to the site. Charges are on the order of $1 to $3 per month and per site.

NOTE

By far, not all premium sites are successful commercially.

E-MAIL

I-Mode services include e-mail. However, if you want, you can also disable e-mail service on i-Mode.

LOOKING AT I-MODE WEB PAGES WITH NETSCAPE OR INTERNET EXPLORER

Of course, because cHTML is an extended subset of HTML, you can use your Netscape or *Internet Explorer* (IE) browser to look at i-Mode pages. Try it out and look at some of the i-Mode pages here: **http://www.eu rotech no logy.com/i/** or **http://www.eu-japan.com/i/**. However,

- At the moment, 99.999 percent of i-Mode users are Japanese and, therefore, almost all i-Mode content is in the Japanese language. Therefore, you will need a Japanese-enabled browser.
- You will not be able to see the links using i-Mode-only tags (such as those tags that dial a mobile voice phone connection directly from the i-Mode handset). Because these special i-Mode tags are not defined in usual HTML, a Netscape or IE browser will not handle them correctly.
- You will not be able to see the many special DoCoMo i-Mode symbols. They will usually be replaced by a question mark.[2]

So looking at an i-Mode page with an ordinary PC-based browser will give you an idea, but will not exactly reproduce what i-Mode users see on their handsets.

WHAT A TYPICAL I-MODE SCREEN LOOKS LIKE

Finally, as previously discussed in Chapter 1, "I-Mode Wireless Fundamentals," Figures 3-1 and 3-2 show what a typical i-Mode screen and handset would look like (i-Mode screen: N502it and a NM502i handset).[3]

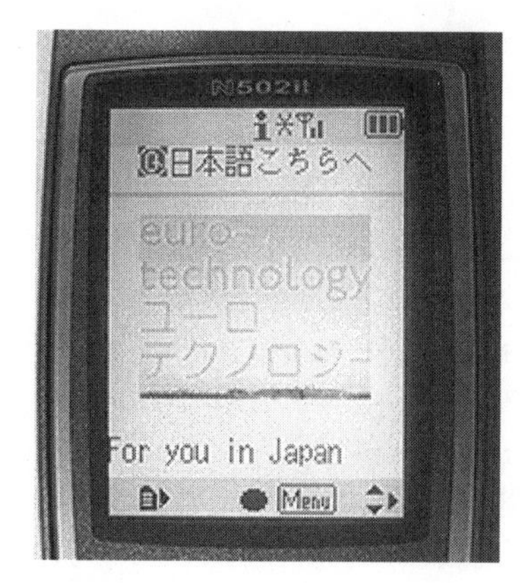

FIGURE 3-1 I-Mode screen (256 colors).

FIGURE 3-2 I-Mode screen (black and white).

NOTE

Around 70 or more different DoCoMo handsets exist (counting color variations) and many more for competing mobile Internet systems in Japan.

CONCLUSION

This chapter showed you how to use i-Mode technology. Today, DoCoMo's i-Mode cell phone service lets subscribers swap e-mail and pictures, search phone directories and restaurant guides, and download news, weather, and horoscopes. Users connect to the Net at 9.6 Kbps, far slower than a PC on a phone modem. However, unlike a PC or other Web-browsing phones, the i-Mode systems are always connected to the Net.

Tomorrow, Internet cruising speeds will soon get a lot faster on i-Mode. More than 300 Kbps should be possible by 2002. By 2003, peak speeds could hit 2Mb, fast enough for high-quality music downloads, Web casts of TV shows, virtual-reality games using the phone as a Net link, and real-time videoconferencing.

END NOTES

[1]John R. Vacca, *Wireless Broadband Networks Handbook: 3G, LMDS, and Wireless Internet*, McGraw-Hill, New York, 2001.

[2]Eurotechnology Japan K. K., Parkwest Building 11th floor, 6-12-1 Nishi-Shinjuku, Shinjuku-ku, Tokyo 160-0023, Japan, 2001.

[3]Ibid.

CHAPTER FOUR

THE TYPES AND SOURCES OF TECHNOLOGY THAT SUPPORT I-MODE

When it comes to mobile commerce, Japan is *light months* ahead of the United States and Europe. Some analysts attribute the phenomenal success of the mobile Internet, in particular the i-Mode service, to the Japanese love of gadgetry.

However, others say this is just a cop-out answer by the European and U.S. carriers who have thus far failed to realize the potential of the mobile Internet. By some estimates, recession-ridden Japan is two years ahead of the rest of the world. So what is the secret of its *mobile* m-commerce success?

KEY FINDINGS: NTT IN THE DRIVER'S SEAT

Already, $600 million worth of m-commerce revenues is generated in Japan annually. For example, as of this writing, NTT DoCoMo's i-Mode service had over 35 million subscribers. This service enables users to send and receive e-mail and gives them access to more than 9,000 Internet sites via their mobile phones.

NTT DoCoMo is the world's second largest mobile phone operator and a subsidiary of telecoms giant NTT. Such is the success of its mobile Internet offering that U.S. Internet giant AOL struck a deal with DoCoMo and yielded control of its Japanese subsidiary to gain access to the i-Mode service.

NTT DoCoMo realized the importance and potential of the mobile Internet. They are the leader to develop this market.

Gadget Love

The high take-up of i-Mode can be attributed to the Japanese love of gadgetry. They (the carriers) took risks based on the assumption that particularly the Japanese would be able to embrace it. They created more powerful applications than is seen in Europe. Once you get critical mass, you get more content. It is a positive spiral.

However, others dismiss cultural factors and say, quite simply, that carriers made the right decisions. The notion that the success of mobile services in Japan is wholly attributable to cultural factors is a handy cop-out by carriers in other regions.

The crucial thing was that NTT and KDDI (another Japanese carrier) offered packet data networks, which makes sure that customers only pay when they send and receive data.

User Freedom

Part of the problem is that carriers outside of Japan try to limit the users' experience to certain sites. They should think about this as growing the market first, not about keeping people contained in a small garden.

Japanese carriers didn't try to hem their users in, as part of a strategy that encouraged content providers to do business with the carriers. Ensuring that the most popular sites got space on the home page, encouraged these content providers to advertise offline, which in turn, boosted the mobile Internet growth.

Paying the Bill

NTT DoCoMo upgraded its billing system so costs for premium services, such as news in English from CNN, can be added to a user's phone bill. Because it was easy to charge for services, content providers were further motivated to tailor their service to suit i-Mode.

Carriers can then make money in two ways: they can charge users for the volume of data transmitted over the packet data network, or they make revenues from services that use their billing systems.

NTT's clear position as leader of a slowly opening market gave it a strong position in talks with manufacturers. It was able to stipulate the design of handsets to hardware vendors. So, although European and U.S. carriers may offer voice handsets with a browser attached, NTT worked with vendors to design handsets optimized for data.

QUIDS IN

For NTT at least, its m-commerce gamble has paid off. Earlier in 2001, when NTT unveiled its results, its bumper profits were linked to the m-commerce boom, in the face of declining profits in a saturated voice telephony market.

In contrast, European carriers have paid hand over fist for *third generation* (3G) mobile phone licenses, which few are convinced will provide a return on their investment. Time is crucial if the rest of the world hopes to catch up with Japan. The year 2002 is crucial for carriers in Europe and the United States to embrace the business models and technology that will make them profitable.

With the aforementioned in mind, new technology-based i-Mode services are now available. What are they? Let's take a look.

I-MODE SERVICES TECHNOLOGY: THE JAVA CONNECTION

Sun Microsystems Inc. recently applauded NTT DoCoMo Inc.'s launch of Java(TM) technology-enabled i-Mode phones and services in Japan. The new i-Mode service, called *Initial*

Applications (i appli), opens the next chapter of i-Mode revolution by providing consumers with new interactive content, including multiplayer games, mobile e-commerce, and in the future, enhanced communications. NTT DoCoMo is Japan's number one mobile phone company and has had phenomenal success to-date with its i-Mode service, signing up more than 29 million subscribers since the service was introduced in February 1999. Now with the launch of Java technology-enabled phones and services, NTT DoCoMo can offer highly differentiated products and services that provide consumers dynamic, personalized, and interactive content.

NTT DoCoMo's launch of its i appli services is a significant event for both Sun and the Java platform. Through Java technology, NTT DoCoMo brings advanced capabilities to the wireless-services marketplace that will enable consumers to take advantage of the many innovative new services being created in Java technology for mobile handsets.[1]

Java technology advances the extraordinary success of i-Mode by improving the capabilities of the next generation of i-Mode handsets and services for customers. Sun's Java technologies and carrier-grade services provide the foundation for NTT DoCoMo to build and expand on their leadership in the wireless market.

The launch of Java technology based i-Mode services represents an important milestone for Java technology. The new 503i cellular phones from Fujitsu and Matsushita (Panasonic) are among the first to be enabled with the Java 2 Platform, Micro Edition™ (J2ME™) technology. Additional i appli handsets incorporating J2ME technology will be provided by other manufacturers in 2002. J2ME technology-enabled interactive services are the next step beyond today's text-based static content. Java software enhances the user experience by supporting easy-to-use, graphical, interactive services for wireless devices. With J2ME technology-enabled phones, users can download software to handsets for disconnected use, as well as access applications interactively from the network.

I appli include a mix of computer games and financial services. Over time, the i appli service will offer new animated games with enhanced graphics and hi-fidelity sound, software for chat, location-based services with zooming maps, secure mobile commerce, and business support programs such as groupware. Already, numerous high-profile content partners have signed up to create new applications and services based on Java technology, including Disney, Bandai, Namco, Sega, and DLJ.

By the way, is that a Web phone in your pocket or are you just glad to speak to me? Actually, i-Mode mobile phones are taking the Japanese directly to cyberspace. Let's take a look.

HANDSETS

Every day, several hundred thousand Japanese switch on the world's only mobile phones with direct links to the Internet to see Hello Kitty (see Figure 4-1).[2] Japanese are queuing to buy i-Mode telephones that give the world's most technologically aware people the ability to download the childlike antics of Asia's most popular cartoon character, the pointy-eared, round-eyed, mouthless (and, above all, cute) Kitty Chan.

Toymaker Bandai Co. and mobile phone operator NTT DoCoMo are gleeful. The Hello Kitty Web site, operated by

FIGURE 4-1 DoCoMo's first color-display i-Mode cell phone. It provides users with direct links to the Internet.

Bandai, is one of the most popular sites on NTT DoCoMo's hugely successful i-Mode mobile Internet service, rapidly growing into Japan's principal platform for e-commerce.

As i-Mode attracts envious glances from Asian and European mobile carriers, it also promises to put Japan (long a laggard in Internet use) at the cutting edge of international cyber culture. A lot people don't mind paying peanuts for such services. Bandai already earns huge revenues.

The i-Mode service owes its popularity to a marriage of Japanese high-tech and pop culture, while enabling easy access to the Net. Just two years after its launch, the service is also one of the few business models for e-commerce to start generating profits so quickly.

The i-Mode keeps users continually linked to the Internet, enabling them to exchange e-mail, swap pictures, call up restaurant guides, and navigate among 7,800 specially formatted Web sites without having to dial up each time. The statistics leave no doubt about its popularity.

Every day a total of 34 million e-mail messages are exchanged by its 20.3 million subscribers. They use the Net an average of 30 times a day, and more than 91 percent pay for some sort of content.

DoCoMo plans to incorporate i-Mode functions into all of its handsets beginning in 2002 and expects i-Mode users to reach 30 million by the end of 2002, matching AOL's current subscriber base of 31 million. DoCoMo will become a large portal service company and plans to develop various e-commerce businesses.

DoCoMo also earns money from e-commerce activity on its platform. It takes a 9-percent commission on billing for its Web site operators, in addition to subscription fees and data transmission fees from subscribers.

I-Mode service brings DoCoMo in additional revenue of 2,000 yen a month per subscriber. That's little more than kitty litter for DoCoMo (the world's second-biggest mobile phone operator with 5.8 trillion yen [$55.73 billion] in annual revenues), but the unique mobile Internet service helped it fend off rivals and dominate Japan's booming mobile phone market.

I-Mode's success has galvanized content providers for mobile communications tools, while spawning a number of agile start-ups, including browser developers. Many are looking to expand their business abroad as Asian and European mobile phone carriers eye their own wireless data services modeled after i-Mode.

In Hong Kong, Hutchison Whampoa Ltd.'s mobile phone arm, in which DoCoMo owns a 19-percent stake, is expected soon to announce plans to adopt the i-Mode service. Inquiries are flooding in from carriers and mobile phone makers in South Korea, Taiwan, and Hong Kong asking Bandai to provide content for their planned mobile Internet services.

The Bandai i-Mode cartoon site that includes Kitty (the pink, mouthless cat with a huge bow on her ear), now generates nearly $33 million in annual revenues from its 1,170,000 members, who pay about $1 a month. Content producer Cybird Co, which offers surfers (in the ocean) up-to-date wave information from around the world, indicates it is in talks with several large Asian carriers on equity tie-ups. It recently sold an equity stake to Intel Corp for an undisclosed sum, as the U.S. chip titan seeks to strengthen its ties with cell phone companies. I-Mode is getting smarter.

As previously explained, DoCoMo plans to incorporate Sun Microsystems Inc.'s Java programming into i-Mode handsets coming out in 2002. That will enable animated figures to move more smoothly or will enable automatic updating of daily news, stock prices, and other information once it has been downloaded. DoCoMo will be able to do many things, like having Kitty Chan moving back and forth across the display, fetching the latest news for you every day.

Now, let's take a detailed look at the latest listing of the best-selling i-Mode-compatible handset models. The ranking in the following list is subject to change without notice.

TOP-SELLING HANDSETS FOR NTT DOCOMO'S I-MODE

The top selling handset is still the P208. However, the just-released P209 handset (see sidebar, "New Internet-Enabled

Handset Suspension for DoCoMo"), which is equipped with a color screen, has already climbed to the second and third position. In fourth place is the P209i as shown in Table 4-1.[3] See Figures 4-2 to 4-17 to view the most currently available i-Mode handsets from NTT DoCoMo. In other words, the company (NTT DoCoMo) is currently offering the handsets shown in Figures 4-2 to 4-17 with i-Mode support.[4]

TABLE 4-1 All Four Top Models Are Manufactured by Matsushita Electric Industries

Ranking	Handset model	Body color
1	P208	lightning silver
2	P209iS	shell pink
3	P209iS	black pearl
4	P209i	crystal metal
5	D209i	misty white
6	N209i	prime gold
7	P209i	crystal white
8	N502i	crescent silver
9	N502i	cincia blue
10	P502i	grace silver
11	P502i	snow white
12	N821i	graphite silver
13	F209i	fine silver
14	N209i	clear pearl
15	D502i	diamond silver
16	P208	lightning black
17	P502i	precious blue
18	SO502i	satin silver
19	SH821i	premium silver
20	D209i	frost pink

NEW INTERNET-ENABLED HANDSET SUSPENSION FOR DOCOMO

NTT DoCoMo recently suffered an embarrassing blow when the companies were forced to suspend sales of their latest Internet-enabled handsets. The Java-capable handsets are made by Matsushita Communications Industrial, the largest mobile handset manufacturer in Japan.

The suspension came after it was discovered that MCI's handset had software glitches, which caused the phones to suddenly lose power when certain Web sites were accessed. Furthermore, DoCoMo indicated, when the phone is switched back on, information, including recorded phone numbers, mail addresses, and saved data, is erased.

The suspension comes a week after DoCoMo announced it would recall 103,000 mobile phone handsets offering its popular i-Mode services, which were found to have software glitches. The handsets, manufactured by Sony, Hitachi Kokusai, and Japan Radio, were found to have minor software problems affecting services, such as e-mail and scheduling.

MCI's Java-enabled handset is one of two models recently launched that enable users to download Java programs for more sophisticated applications. MCI's handset is the more popular of the two, with 341,000 already sold, compared with 77,000 for the other model.

DoCoMo reacted swiftly to the problem, which is believed to stem from a software bug. The problem happens very rarely.

Nevertheless, the suspension comes at a sensitive time for DoCoMo and MCI, which are preparing for the launch of advanced 3G mobile phones in 2002. Analysts are increasingly concerned that manufacturers will not be able to deliver their handsets on time due to the complexity of 3G handsets.

P210i
Manufacturer: Panasonic (Matsushita)
Height: 120mm
Width: 40mm
Thickness: 14mm
Weight: about 59g
Talking time: 140 min
Standby time: 400 hrs
Colors: Blackberry, Silverberry, Leaf

D210i
Manufacturer: Mitsubishi
Height: 123mm
Width: 41mm
Thickness: 17mm
Weight: about 71g
Talking time: 125 min
Standby time: 500 hrs
Colors: Brilliant White, "Kirakira" Pink, Sparkling Silver

FIGURE 4-2 Handsets: P210i and D210i.

F210i
Manufacturer: Fujitsu
Height: 125mm
Width: 40mm
Thickness: 15mm
Weight: about 65g
Talking time: 135 min
Standby time: 500 hrs
Colors: Cool Silver, Happy Orange, Cyber Grey

N210i
Manufacturer: NEC
Height: 90mm
Width: 46mm
Thickness: 23mm
Weight: about 98g
Talking time: 135 min
Standby time: 500 hrs
Colors: Siphon Purple, Aquatic Blue, Glamourous Silver

FIGURE 4-3 Handsets: F210i and N210i.

P503is
Manufacturer: Panasonic (Matsushita)
Height: 97mm
Width: 50mm
Thickness: 27mm
Weight: about 98g
Talking time: 145 min
Standby time: 440 hrs
Colors: Starling Silver, Jet Black, Ocean Blue

SO503i
Manufacturer: Sony
Height: 98mm
Width: 49mm
Thickness: 28mm
Weight: about 115g
Talking time: 140 min
Standby time: 210 hrs
Colors: White&Silver&Black, Candy

FIGURE 4-4 Handsets: P503is and SO503i.

D503i
Manufacturer: Mitsubishi
Height: 129mm
Width: 46mm
Thickness: 17mm
Weight: about 81g
Talking time: 130 min
Standby time: 380 hrs
Colors: Lightning White, Metallic Rouge, Blue Graphite

N503i
Manufacturer: NEC
Height: 93mm
Width: 48mm
Thickness: 22mm
Weight: about 98g
Talking time: 135 min
Standby time: 460 hrs
Colors: Airy, Indigo, Lavendel

FIGURE 4-5 Handsets: D503i and N503i.

P503i
Manufacturer: Panasonic (Matsushita)
Height: 128mm
Width: 45mm
Thickness: 17mm
Weight: about 74g
Talking time: 140 min
Standby time: 400 hrs
Colors: Silver, Gem Black, Cyber Red

F503i
Manufacturer: Fujitsu
Height: 135mm
Width: 46mm
Thickness: 15mm
Weight: about 77g
Talking time: 135 min
Standby time: 430 hrs
Colors: Fairy White, Silver, Shine Grey

FIGURE 4-6 Handsets: P503i and F503i.

R691i
Manufacturer: Japan Radio
Height: 130mm
Width: 48mm
Thickness: 20mm
Weight: about 99g
Talking time: 120 min
Standby time: 430 hrs
Colors: Active Orange, Navy Blue
Waterproof Outdoor model

ER209i
Manufacturer: Ericsson
Height: 106mm
Width: 50mm
Thickness: 18mm
Weight: about 77g
Talking time: 130 min
Standby time: 310 hrs
Colors: Aurora Blue
Bilingual / English operation manual available

FIGURE 4-7 Handsets: R691i and ER209i.

KO209i
Manufacturer: Kokusai (Hitachi)
Height: 125mm
Width: 41mm
Thickness: 15mm
Weight: about 69g
Talking time: 120 min
Standby time: 350 hrs
Colors: Trashy White, Brilliant Silver

SO502iWM
Manufacturer: Sony
Height: 105mm
Width: 50mm
Thickness: 28mm
Weight: about 120g
Talking time: 100 min
Standby time: 200 hrs
ATRAC 3 Music Playback

FIGURE 4-8 Handsets: KO209i and SO502iWM.

R209i
Manufacturer: Japan Radio
Height; 123mm
Width: 39mm
Thickness: 15mm
Weight: about 63g
Talking time: 120 min
Standby time: 430 hrs
Colors: Cosmo Silver, Aurora White

N502it
Manufacturer: NEC
Height: 93mm
Width: 48mm
Thickness: 22mm
Weight: about 105g
Talking time: 130 min
Standby time: 460 hrs
Colors: Creamy White, Titan Black, Moonlight Silver

FIGURE 4-9 Handsets: R209i and N502it.

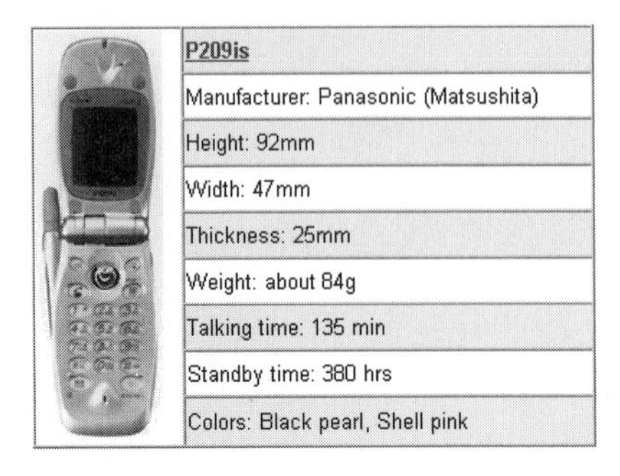

P209is
Manufacturer: Panasonic (Matsushita)
Height: 92mm
Width: 47mm
Thickness: 25mm
Weight: about 84g
Talking time: 135 min
Standby time: 380 hrs
Colors: Black pearl, Shell pink

F209i
Manufacturer: Fujitsu
Height: 125mm
Width: 40mm
Thickness: 15mm
Weight: about 63g
Talking time: 135 min
Standby time: 450 hrs
Colors: Airy blue, fine silver
Bilingual / English operation manual available

FIGURE 4-10 Handsets: P209is and F209i.

N209i
Manufacturer: NEC
Height: 90mm
Width: 46mm
Thickness: 19mm
Weight: about 86g
Talking time: 120 min
Standby time: 500 hrs
Colors: Prime Gold, Clear Pearl, Sweet Pink
Bilingual / English operation manual available

P209i
Manufacturer: Panasonic (Matsushita)
Height: 123mm
Width: 39mm
Thickness: 15mm
Weight: about 60g
Talking time: 135 min
Standby time: 350 hrs
Colors: Crystal White, Crystal Metal, Vintage
Bilingual / English operation manual available

FIGURE 4-11 Handsets: N209i and P209i.

D209i
Manufacturer: Mitsubishi
Height: 124mm
Width: 41mm
Thickness: 18mm
Weight: about 74g
Talking time: 120 min
Standby time: 400 hrs
Colors: Misty White, Dark Titan, Frost Pink
Bilingual / English operation manual available

SO502i
Manufacturer: Sony
Height: 122mm
Width: 42mm
Thickness: 17mm
Weight: about 73g
Talking time: 120 min
Standby time: 210 hrs
Colors: Satin Silver, Charcoal Grey, Platina Blue

FIGURE 4-12 Handsets: D209i and SO502i.

F502it
Manufacturer: Fujitsu
Height: 125mm
Width: 42mm
Thickness: 19mm
Weight: about 71g
Talking time: 130 min
Standby time: 340 hrs
Colors: Premium Gold

NM502i
Manufacturer: Nokia
Height: 111mm
Width: 44mm
Thickness: 18mm
Weight: about 77g
Talking time: 130 min
Standby time: 270 hrs
Colors: Silver Putty, Blue Velvet
Bilingual / English operation manual available

FIGURE 4-13 Handsets: F502it and NM502i.

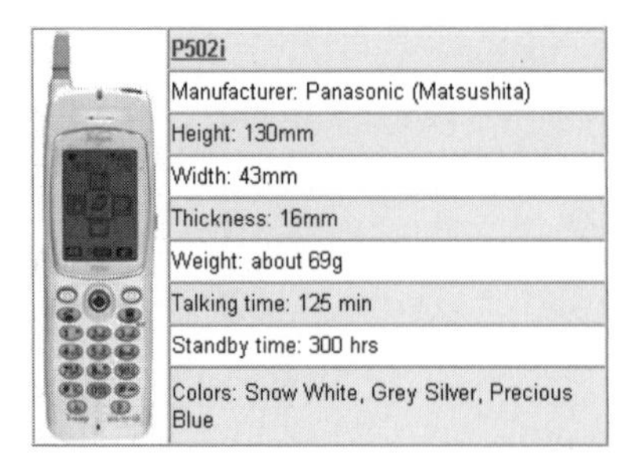

P502i
Manufacturer: Panasonic (Matsushita)
Height: 130mm
Width: 43mm
Thickness: 16mm
Weight: about 69g
Talking time: 125 min
Standby time: 300 hrs
Colors: Snow White, Grey Silver, Precious Blue

N502i
Manufacturer: NEC
Height: 93mm
Width: 48mm
Thickness: 22mm
Weight: about 98g
Talking time: 120 min
Standby time: 420 hrs
Colors: Cherry Blossom, Crescendo Silver, Sincere Blue

FIGURE 4-14 Handsets: P502i and N502i.

D502i
Manufacturer: Mitsubishi
Height: 132mm
Width: 43mm
Thickness: 20mm
Weight: about 84g
Talking time: 130 min
Standby time: 350 hrs
Colors: Metallic Rose, Lilac Pearl, Pure White, Diamond Silver

F502i
Manufacturer: Fujitsu
Height: 125mm
Width: 40mm
Thickness: 19mm
Weight: about 71g
Talking time: 130 min
Standby time: 340 hrs
Colors: Silky White, Prime Silver

FIGURE 4-15 Handsets: D502i and F502i.

P821i (Doccimo)
Manufacturer: Panasonic (Matsushita)
Height: 129mm
Width: 41mm
Thickness: 19mm
Weight: about 82g
Talking time: 115 min (Keitai), 390 min (PHS)
Standby time: 250 hrs (Keitai), 340 hrs (PHS)
Colors: Dark Metallic, Noble Silver

N821i (Doccimo)
Manufacturer: NEC
Height: 93mm
Width: 48mm
Thickness: 24mm
Weight: about 105g
Talking time: 120 min (Keitai), 420 min (PHS)
Standby time: 380 hrs (Keitai), 600 hrs (PHS)
Colors: Graphite Silver

FIGURE 4-16 Handsets: P821i (Doccimo) and N821i (Doccimo).

SH821i (Doccimo)
Manufacturer: Sharp
Height: 126mm
Width: 39mm
Thickness: 20mm
Weight: about 76g
Talking time: 110 min (Keitai), 450 min (PHS)
Standby time: 270 hrs (Keitai), 400 hrs (PHS)
Colors: Premium White, Premium Silver

FIGURE 4-17 SH821i (Doccimo)

HANDSET SOFTWARE

NTT DoCoMo Inc. is slated to repair about 200,000 Web-enabled cell phones, the latest in a series of recent technical snafus to dog Japan's cutting-edge mobile services. DoCoMo said it would also temporarily stop sales of the Panasonic phones, made by *Matsushita Communication Industrial Co Ltd.* (MCI), after they found that they were unable to receive incoming calls at certain geographical locations because of continued handset software glitches.

This was the second technical glitch in 2001 to hit the Panasonic P503i Hyper series, which connects to DoCoMo's popular i-Mode's Internet service and uses the Java computer language to enable games and other advanced applications. Launched in January 2001, the P503i Hyper has become a standard part of DoCoMo's product lineup with more than 900,000 units sold.

Japan's handset makers have been hit by a series of costly recalls recently as they struggle with increasingly complex software and technology for the sophisticated functions and Internet connections that have put Japan at the forefront of global mobile technology. As handsets become more advanced, more and more points need to be checked and many things cannot be anticipated. The impact of the latest incident, if any, would be primarily psychological.

Browsers

It's another blow for sentiment, but in reality, in terms of the numbers, it's a minor incident. It affects only about 2.5 percent of Java-enabled browser handsets out there (see Table 4-2).[5]

Investors will be upset. It certainly doesn't help Matsushita. DoCoMo is probably quite upset with them.

In Japan, where carriers work very closely with their vendors, DoCoMo and Matsushita, Japan's biggest handset browser maker, have collaborated closely on product development and research (see Table 4-3).[6]

The series of setbacks for Java and Internet-enabled phones deals another blow to Japan's reputation as a leader in wireless technology. It was smudged when DoCoMo delayed by months, until 2002, the commercial launch of 3G mobile services that will offer videoconferencing and other futuristic functions. That is still set, however, to be the world's first commercial 3G launch.

TABLE 4-2 I-Mode Browsers and Emulators

TYPE/COMPANY	PRODUCT DESCRIPTION
Access (Japan) Access (USA)	Based on a subset of HTML 4.0 standard, Compact to NetFront extends the existing Internet infrastructure support the next generation of mobile information appliances with memory, display, and bandwidth constraints. With NetFront embedded in cellular phones, users can access various services, such as weather forecasts, transportation schedules, data searches, and news updates, at anytime and anywhere.
C.media	C.media i-Mode simulator.
Ezos	cHTML/WAP/xHTML browser.
i-browser	I-browser i-Mode browser for Windows.
i-Monde	I-Mode browser.
iEmulator	Java i-Mode emulator.
Seiichi Nishimura	ITool i-Mode Emulator: Emulates D501i/F501i, P501i, and N501i (in Japanese).
NTT DoCoMo	I-Mode World i-Mode demo.
OpenWave (formerly phone.com+software.com)	I-Mode support announced Jan. 2001.
Pixo	Pixo Internet microbrowser. Supports standard HTML and cHTML.
X9	I-Mimic: i-mode emulator.
Zentec	I-JADE: A platform that emulates an i-Mode cell phone with built-in KVM on an ordinary PC allowing the user to debug applications during development for target cell phones and retain the Java development capital.

DoCoMo gave no estimates for the cost of the repairs and had yet to determine who would pay for them, although such costs are usually shouldered by the manufacturers. DoCoMo indicated that the problems, which sometimes prevented outside calls from going through, were caused by a design flaw in the phone's microprocessor (developed by MCI). Nevertheless, the glitch could be fixed without replacing the faulty part.

TABLE 4-3 I-Mode Editors and Development Tools

Company	Product description
Adobe	**GoLive** Adobe GoLive goes wireless (PDF) (i-Mode support announced Q3 2000).
AVIDWireless	**AVIDRapidTools** A Java toolset that allows development of device and format-independent wireless applications. ART is a set of reusable Java beans that provides the architecture for handling multiple sessions, personalization, and data access, freeing the developer to concentrate on the business needs of the wireless application. ART supports multiple formats including: WAP/WML, i-Mode/cHTML, HTML, Palm VII/PQA.
EpocketCash	The secure and anonymous payment system.
IBM	**WebSphere studio homepage for Windows** With everything needed to design, build, and publish your Web pages. Now includes i-Mode support to create Web pages for cellular phones and PDAs.
Mercury Interactive	**LoadRunner** Load-testing software and ActiveTest, a hosted version, that can be used to monitor Web applications based on WAP and i-Mode.
Network365	**mZone** Mobile commerce server for WAP, i-Mode, and SMS.
Nooper	**NooperLabs** A testing service for your i-Mode application. NooperLabs is an i-Mode/i appli application testing service. It provides your business with a means to test your application on a variety of real i-Mode handsets.
NTT DoCoMo	I-Mode Java pages (in Japanese).
Sanface	**PDFmail** Send PDF files via an i-Mode phone.
wapprofit.com	**wapprofit i-Mode editor** The wapprofit i-Mode editor is an easy to use product for Windows 95/98, Windows NT, and Windows 2000 systems. It simplifies the development of i-Mode pages and sites. It supports the insertion of special i-Mode tags, auto-creation of i-Mode menus with phone access keys, cross-referenced, and enhanced display properties. It also has a preview that emulates the performance of an i-Mode handset.
Woolox	The Woolox chart library is a collection of custom *JavaServer Pages*™ (JSP™) 1.1 tags and a class library for the dynamic chart generation. It can produce several chart images to be used in HTML and WML pages (WAP use).

Company	Product description
	Woolox 2D is a toolbox for the creation of dynamic images. The library is a collection of custom JSP tags and a class library for the dynamic image generation. It can produce images to be used in HTML, cHTML (i-Mode), and WML pages (WAP). The library includes server-side JSP tags for easy image production. No Java programming skills are needed.
X-Servlet from Flex Firm.	X-Servlet is a state-of-the-art mobile application server. X-Servlet is deployed worldwide on the existing servers of content providers, aggregators, and mobile carriers. It distributes sophisticated media content to mobile devices such as WAP phones, PDAs, pagers, i-Mode devices, Car-Internet, Digital TV, and anyone in the future such as WAP-NG.
XIAM	The XIAM information router allows for the intelligent routing of information between an enterprise and mobile users over SMS. Remote users can easily update a database, reply to e-mails, access an intranet or Internet Web page, or even restart a critical program from any standard mobile device at any time.
Xybo	Applications built using Xybo wireless application builder work with virtually all wireless standards and devices including WAP phones, i-Mode terminals, two-way pagers, PDAs, and so on.
Zentec	**I-JADE** A complete development environment for i-Mode-compliant Java applications.

Recently, Sony Corp announced its third recall of browser handsets in Japan (estimating the series of slip-ups)—which in some cases, required replacing the entire phone (which would cost it an estimated 12 billion yen [$96 million]). In addition, MCI's shares recently managed a modest bounce with the rest of the Japanese high-tech sector, ending 1.63 percent higher at 5,000 yen after slipping to a 33-month low. The company's shares have been hit hard by a persistent slump in demand for cell phones. NTT DoCoMo's shares ended up 1.47 percent at 2.07 million yen.

Business Software

Keeping the aforementioned in mind, the GSM Association[7] (the world's leading wireless industry body), recently announced the creation of the *Mobile Services Initiative* (M-Services), an unprecedented global industry move to enhance benefits to consumers using GSM and the new i-Mode handsets by delivering a globally available set of services and business software through the mobile Internet. The Association has acted to mobilize the wireless industry, bringing together its operator community to provide clear guidance to i-Mode handset manufacturers and business software developers on the needs of consumers of mobile Internet services going forward.

The wireless industry has established a clear need for M-Services, and working with leading manufacturers of handsets and service platforms, the Association has compiled a set of prioritized, compelling service guidelines (on a freely available basis), that can be offered in the future by any manufacturer of *General Packet Radio Service* (GPRS) handsets. The M-Services initiative was undertaken by the GSM Association to enable GPRS users to experience a new level of consistently available services through the mobile Internet. These services could include enhanced graphics, music, video, games, ring tones, screen savers, and other compelling services, which will be brought easily to the mobile screens of consumers.

The Association's initiative received ringing endorsements from all sectors of the wireless industry. This landmark initiative is a major step towards delivering compelling, universally available services to consumers worldwide. M-Services is a big step forward in making the mobile Internet work, for consumers, service providers, and the entire mobile Internet ecosystem.

Manufacturers, such as Alcatel, Ericsson, Motorola, Nokia, Sagem, Samsung, and Siemens, have quickly come out in support of the M-Services initiative. Capitalizing on widely adopted standards, such as *Short Message Service* (SMS) and EMS, M-Services is a major milestone helping operators bring truly compelling services to end-users.

Ericsson supports the M-Services initiative as an important vehicle to ensure volume use of mobile Internet in a standardized and non-proprietary way. M-Services will leverage other key standardization efforts, like WAP, EMS, MMS, SyncML, and the new i-Mode, to bring a consistent user experience for digital content.

M-Services will demonstrate the power of GPRS by delivering an enriched mobile experience to consumers, and it should go a long way towards accelerating the adoption of wireless data. The development of open standards is essential to the interoperability between services and devices.

SMS has shown the immense value of offering a universally available service to mobile users. Already running at a level of 29-billion messages a month, it is expected that consumers will send over 300-billion text messages in total across GSM networks globally in 2002.

The GSM Association, in recognizing the value of universally available key services, moved quickly to identify and offer to handset manufacturers the ability to enable these services through their GPRS handsets. It's inevitable that the manufacturers will do all they can to implement M-Services into their handset product offerings in a prioritized, quality way.

France Telecom Mobiles also recognized the GSM Association's timely, industry-leading efforts in bringing out these services. France Telecom Mobiles views M-Services as very positive and timely in enhancing the consumer's experience in connection with newly available GPRS services.

Telefónica Móviles believes that the M-Services initiative will enable the entire GSM/GPRS mobile industry to demonstrate the enormous potential of this technology to the world. Telefónica Móviles encourages all GSM/GPRS operators to support M-Services and thereby create a consistent basis on which to build their service offer for clients. In doing so, they as operators, will obtain handsets with the most advanced functionalities for their consumers as quickly as possible.

In effect, M-Services will take all of us into the third dimension (voice, data, and services) before 3GSM arrives. As a part of BT Wireless, Genie is already offering consumers a wide variety of compelling services. The M-Services initiative will leverage and enhance these service offerings in the current environment and take all of us through to GPRS and 3GSM. M-Services builds on the good work of the standards bodies.

The SMS success story points up the value of making such services universally available. GPRS signals the true arrival of the mobile Internet. This initiative will help to positively and consistently position this evolution as a real benefit to customers rather than some over-hyped technology that will fail to live up to expectations.

M-Services is seen as setting operator requirements. The introduction of M-Services will be a significant boost to the attraction of GPRS, taking the focus off just speed and on to the compelling nature of services and content—the real drivers of the mobile Internet.

Nevertheless, the mobile Internet, with its new technologies and services, is a tough concept to grasp. Having plunged deeply into NTT DoCoMo's i-Mode universe, how fast is i-Mode?

Home Page Compression Technology

The current i-Mode service has a transmission speed of 9.6 Kbps, which is like using a steam locomotive in today's world. However, it's good enough if you're using a packet-switching system, as the i-Mode service does. You're not going to get full-motion video, but you can download images at a speed of two frames per second. E-mail is limited to about 250 characters per message, but you can get a series and scroll down. The current screen home page compression technology leaves a lot to be desired, though, and reading these messages can strain your eyes. The i-Mode service uses compression technology to increase the volume of data that's transmitted, and the underlying packet-switching system makes efficient use of bandwidth.

So what will the extra speed let you do? If you want to order a video for the evening, you'll be able to call the rental outfit and download clips to your smart phone while riding the train home. When you get home, you'll stick the handset in a jack that's connected to your home entertainment system and download the movie. Go get a glass of wine, and when you return, you can sit down and watch that film. Think of what you're doing now with the desktop Internet and transfer it to the mobile Net, and remember: the handsets, display, compression, and other technologies are only going to improve.

So, what are the technology services that support I-Mode? Let's take a look.

TECHNOLOGY SERVICES THAT SUPPORT I-MODE

Recently, search engine vendor Google Inc. indicated that it has released wireless search technology services for users of i-Mode wireless Net-ready phones. The technology, released in Japan, enables i-Mode users to search and browse more than 2.4 billion Web pages. When a user finds a traditional HTML Web page, the technology automatically translates it into a format optimized for i-Mode. The search engine is available at **http://www.google.com/imode**.

I-MODE HOME PAGES

In addition to its own i-Mode home page search site, Google Inc. also provides a variety of Web search technology to OEMs and other vendors. It claims its technology serves more than 70 million searches a day.

CONTENT PROVIDER FEES

The overall pattern of development and use of payment systems and electronic commerce in Japan is not dissimilar to that in Europe: *electronic money* (e-money) schemes are stagnating, and the majority of payments in e-commerce is made using payment methods also favored in the bricks-and-mortar world. The costs of Internet use are a major barrier to the rapid growth of e-commerce as is reluctance to transmit credit card information across the Net. New developments frequently involve mobile telephones.

Though Internet users in Japan number almost 40 million, there is no booming market in e-commerce. The volume of e-commerce in Japan hit $5 billion in 2001, and is still growing, but remains only 20 percent of the U.S. *e-Christmas* market in 2000. The high cost of communication lines is said to be the main reason for that, because it takes time to find the desired goods by browsing Web pages. Restrictions in payments over the Internet probably are also a barrier.

After the eruption of the Internet in Japan in 1994, several e-payment trials supporting e-commerce were introduced. Those are e-money payment, credit card payment, and debit card payment. Though debit card payment is definitely a kind of electronic payment, it means that because it also uses communication lines, this method is only used at *bricks and mortar* shops.

At the early stage, e-money seemed the most promising payment system, and several different kinds of e-money systems were experimentally provided, following Mondex in Swindon in the United Kingdom. These included Visa cash, Internet cash, and super cash. The last two were invented and developed by *Nippon Telegraph and Telephone Corporation* (NTT). However, e-money systems shrank very rapidly after the termination of the experiments, not because of technical issues, but because of the complicated registration procedures that involved opening an extra account at the bank and receiving a special IC card holding e-wallet-handling software in addition to an ATM card.

At most, only hundreds of thousands of IC cards were delivered. Of course, this is one of the reasons that e-money has not become a popular payment system. If compared with the hundreds million of credit cards in circulation throughout Japan, which have established their position as a convenient payment system, the result with respect to competitiveness was clear. People did not want to acquire another payment card for use only over the Internet. What's more, only a few million home users could access the Internet through telephone lines or cable lines, where the transmission rate is not so good.[8] For example, though ISDN users can enjoy 64 Kbps and CATV users 1.5 Mbps, people using normal dial-up systems can only use 36 Kbps at most. This low bit rate of access to the Internet and the high cost for communication lines discouraged users from purchasing goods at Web shops. Though users were not required to pay any e-money handling costs and did not need to go to the shop, they sometimes paid more than the conventional purchase price at shops because of the high cost for the lines caused by both rate and time spent for purchase.

Consequently, it seems very natural for credit card payment to become popular. It is a fact that the credit card payment system is the most prevalent form of electronic payment. Users, however, often prefer the remittance system through banks or post offices after they have confirmed the quantity and quality of goods they want to purchase at Web shops, just as they would do by browsing catalogs sent by vendors. They do not use the e-payment system, though they use the Internet, as an accessing path to Web shops. It is said that Japanese customers do not want to send their credit card ID numbers through communication lines even though they know the information is securely transferred to the partner by SSL or SET.

The total amount of trading over the Internet in 2001 was 572 billion yen ($5.4 billion), more than twice as much as the previous year 2000, but this is less than 0.1 percent of annual expenditure by customers in Japan. The average of each transaction is about 15,000 yen ($140). Not all transactions are settled by credit card—indeed credit card payment over the Internet has not achieved popularity.

Is there any other booming payment system in Japan? Indeed, there is: the J-Debit service is now growing rapidly. More than 500 million ATM cards held by customers can be used in the J-Debit service using about 370,000 J-Debit terminals located in shops and branches of big retailers. Total trade for the first half-year surpassed $1.2 billion with 4.6 million transactions. It is not so clear why Japanese customers prefer this payment compared with credit card payment. However, it can be said that availability in convenience stores, discount sellers, in addition to department stores and gas stations, as well as participation and cooperation by major banks and local trusted banks, are the main reasons that this system has a good reputation.

The i-Mode payment system that has recently become available consists of two types: one is to enable NTT DoCoMo to collect the fee for content that the user buys at an official site. (An official site is a Web site where a content provider opens a virtual shop to sell a variety of contents, like signaling melodies or images for waiting display.) This seems similar to the *900 number* service in the United States and is called as Dial Q2 service in Japan.

In this service, the upper limit is set to 300 yen ($3) a month, because NTT DoCoMo assumes the risk of having to pay the fee for the goods to the vendor if NTT DoCoMo cannot collect the fee from users. NTT, DoCoMo's parent company, had a bitter experience in collecting fees for the Dial Q2 service.

The other system is that a user pays the fee for e-trade goods with his or her credit card ID. The users send their credit IDs through the i-Mode browser just as they do through a PC. In this case, users can buy goods more expensive than the digital goods mentioned in the proceeding.

Applications include telephone banking (payment of bills or remittance), telephone e-trade for stocks, or e-registration for concert tickets or train/flight seats. As is seen by these examples, trading acts themselves are made over the Internet, but payment acts are not necessarily processed via the same channel.

The latest news is that some more sophisticated payment methods are now being developed at NTT DoCoMo. DoCoMo announced that e-money charged to IC cards (probably of the wireless type) through cellular phones will be available soon. This e-money can be used in both real and virtual shops. E-tickets for soccer games held in cellular phones after payment are completed by charging through the Internet and can be used to enter the stadium just by showing the ticket on the display.

Thus, many ideas for e-commerce and e-payment have been created and tested by developing prototype systems.[9] Some of them have become very popular, supported by emerging communication tools like low-cost PCs or cellular phones with i-Mode. It will be interesting to keep track of these technical trends in hardware and of the applications made available on this hardware.

On another note, DoCoMo has been asked by the Japanese government to loosen control over the i-Mode network, search engines and portals. Let's see why.

I-MODE-LIKE SERVICE

NTT DoCoMo recently announced that it will increase its stake in *KG Telecommunications Co., Ltd.* (KG Telecom), a Taiwanese GSM cellular phone operator, by purchasing a maximum of 62,360,884 new shares for existing shareholders and employees. KG Telecom will issue a total of 187,332,096 new shares at NT $30 per share, meaning NTT DoCoMo's purchase could be worth up to approximately NT $1.87 billion (about 6.92 billion yen).

NTT DoCoMo will acquire the shares through DCM Capital TWN (UK) Ltd. and Taiwan DoCoMo Ltd., which were established as a result of NTT DoCoMo's acquisition of KG Telecom stock on November 30, 2000. Koos Group, the largest shareholder of KG Telecom, and NTT DoCoMo, will be entitled to purchase the unsubscribed shares up to the ratio of their respective holdings in the company.

For NTT DoCoMo, which provides KG Telecom with technological support and is seating two members on its board of directors, the increased investment will pave the way for the early deployment of an i-Mode-like and 3G wireless services in Taiwan, as well as enhance the value of KG Telcom.

NTT DoCoMo and KG Telecom will enter into a licensing agreement for the introduction of i-Mode-like services in Taiwan, targeted at around mid-2002. NTT DoCoMo will license technology and provide know-how to introduce the service over a GPRS network for i-Mode HTML/WML dual-browser handsets. It will also provide consulting for a business plan feasibility study, server and terminal development, PR and marketing, portal management, content development, and customer care. NTT DoCoMo will receive fees from KG Telecom for the license and services. [10]

SEARCH ENGINES AND PORTALS

NTT DoCoMo has been asked by the Japanese government to open parts of its i-Mode network and services to third-party companies (see sidebar, "I-Mode-Like Service"). At present, DoCoMo controls the entire process of the i-Mode service, including the gateways that enable users to access content provider sites. DoCoMo also operates the only search engines and portals for i-Mode that make up the official i-Mode menu on phones issued by DoCoMo.

However, the Ministry of Public Management, Home Affairs, Post, and Telecommunications, representing the ISPs and content providers, has asked DoCoMo to enable third-party operators to set up independent gateways and portals. The ministry will consider introducing regulations to force open the i-Mode service if DoCoMo doesn't comply with the request, AsiaBizTech reported.

CONCLUSION

This chapter showed you the types and sources of technology that support i-Mode. In conclusion, future mobile phones will not be as different from current ones as had been thought. In other words, future generation mobile phones, for which operators have spent billions of dollars securing airwave space, will not prove the cash cows firms had hoped.

Japanese phone operator NTT DoCoMo, which pioneered Internet access from mobile phones, believes that 3G handsets will be little advanced from previous models. Although 3G phones will be able, as operators hope, to receive color video transmissions and high-quality music, large files will prove too costly to download. The conclusion is that NTT DoCoMo will perhaps offer short video clips of 10 to 15 seconds and previews of music that people can purchase to download at home through their PCs or TVs.

The ability of 3G to carry video and sound clips has been seen as a large potential revenue stream for operators. The technology gap, which NTT DoCoMo has been researching, will leave firms struggling to justify the 100 billion Euros spent on buying European licenses to operate the services.

The business model probably will not fundamentally change from 2G to 3G. The essence of the cellular phone business will be the same.

Finally, there are increasing concerns over the sums paid by operators earlier in 2001 for airwaves used for running 3G licenses. The UK and German governments netted $35 billion (£22bn) and $46.1 billion respectively.

However, more recent auctions have been less buoyant, with Italy gaining less than half the money it had hoped for. Switzerland recently postponing its 3G auction after the number of bidding operators matched the number of licenses on offer. The Swiss Federal Communications Office is expected to announce its next move in 2002.

END NOTES

[1]NTT DoCoMo, Inc., Sanno Park Tower from 27F to 44F, 11-1 Nagatacho-2-chrome, Chiyoda, Tokyo, Japan, 2001.

[2]John R. Vacca, *Wireless Broadband Networks Handbook: 3G, LMDS, and Wireless Internet*, McGraw-Hill, New York, 2001.

[3]"NTT DoCoMo's I-Mode Handsets," WestCyber Corporation, 1350 Broadway, Rm. 705, New York, NY 10123, 2001.

[4]Ibid.

[5]palowireless.com, 1 Bogan Street, Summer Hill NSW 2130, Australia, 2001.

[6]Ibid.

[7]GSM Association, Avoca Court, Temple Road, Blackrock, Co. Dublin, Ireland, 2001.

[8]John R. Vacca, *The Cabling Handbook* (2nd Edition), Prentice Hall, Upper Saddle River, N.J., 2001.

[9]John R. Vacca, *Electronic Commerce: Online Ordering and Digital Money with CD-ROM*, Charles River Media, 2001.

[10]"DoCoMo To Increase Stake in KG Telecom, License Technology for I-Mode-Like Service," AnywhereYouGo.com, 3000 Waterview Parkway, B2E14, Richardson, TX 75080, 2001.

PART II

DEVELOPING I-MODE APPLICATIONS

CHAPTER FIVE

OFFICIAL I-MODE MENU SITES

I-Mode sites can be divided into two basic types: official i-Mode sites, and unofficial or voluntary sites. Official i-Mode sites are ones that appear automatically on the i-Menu of any i-Mode mobile phone, because they have been officially checked, approved, and listed there by NTT DoCoMo. Unofficial sites are not listed on the i-Menu, but can be reached by typing in the URL or sending a bookmark to the phone by e-mail. These sites have no official connection to NTT DoCoMo's i-Mode service.

In other words, the i-Mode platform, like the AU and J-Sky platforms, mainly offer e-mail and Web sites browsing services. Official Web sites and nonofficial Web sites exist. To be listed on the i-Mode menu and be accessible by a simple click, content providers have to obtain the agreement of NTT DoCoMo (which runs the i-Mode platform). The nonofficial Web sites are designed by private individuals and companies. The difference is that they are not controlled neither listed on the i-Mode menu. An estimated 3,000 official Web sites and 40,000 nonofficial Web sites exist. Both categories provides entertainment, chat, news, travel info, shopping information, culture contents, adult contents, and so on. It is possible to find directories and search engines listing all the i-Mode Web sites. Some of the sites are also accessible from the Net.

As previously stated, over 40,000 i-Mode-compatible Web sites exist, of which about 3,000 are official sites. Some of them are free, whereas others charge monthly fees ranging from 100 to 300 yen. The official Japanese i-Mode sites are categorized into

- News and information
- Mobile banking
- Charge cards, stocks, and insurance
- Travel
- Ticket and living
- Gourmet and recipes
- Entertainment
- Town information
- Dictionary and convenient tools[1]

NOTE

The majority of i-Mode sites are in Japanese only! You can find a lineup list of official sites without active links on the DoCoMo Web site: ***http://www.nttdocomo.co.jp/i/corp/teikyou.html*** *along with a short service description and an indication of whether the service is free or fee-based. A good starting point to find Japanese-language unofficial i-Mode sites is the Infoseek Japan i-Mode directory at* ***http://www.infoseek.co.jp/Topic/3500/3588 /3572*** *or at* ***http://www.ohnew.co.jp/****.*

ACCESSING I-MODE SITES

On your i-Mode handset, after users press the *i-Mode button* and the *i-Menu* soft-button/link, a new menu appears that includes an entry for the *menu list*. As previously stated, those i-Mode sites, which users can access by clicking on links on the *menu list* are *official* i-Mode sites. Those i-Mode sites that are not directly linked to sub pages of the *menu list* are *unofficial* (katte) sites.

FINDING AND ACCESSING ENGLISH-LANGUAGE I-MODE SITES

I-Mode doesn't have many English-language sites yet, which is mainly due to the fact that i-Mode is available only in Japan so far. The official English-language i-Mode menu from NTT DoCoMo can be found on the phone menu under *i-Menu* → English → Menu List as shown in Figure 5-1.[2]

NOTE

*DoCoMo also offers a lineup list without active links on the DoCoMo Web site (**http://www.nttdocomo.com/i/imenu/index.html**).*

The i-Mode-accessible portion of the i-Mode links Web site (this Web site) provides a categorized guide to the growing number of both official and unofficial English-language i-Mode sites. The sites listed in the guide are categorized as follows:

- News and sports
- City data (town guides and event listings for Tokyo and elsewhere)
- Services, info (bank information, online dictionaries, and so on)

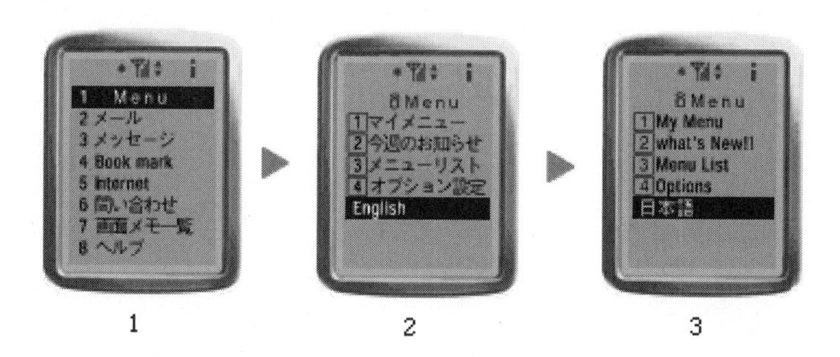

FIGURE 5-1 Official English-language i-Mode menu.

- Amusements (games, ring tones, and light reading)
- Japanese links (selected useful information and services not yet available in English)[3]

WHY ENGLISH SITES LOOK SO STRANGE ON I-MODE English sites look so strange on i-Mode, because i-Mode was designed for use in Japan. I-Mode phones and their built-in microbrowsers are made to display Japanese text, which doesn't require spaces between words. Desktop Web browsers, like Netscape Communicator and Internet Explorer, automatically insert line breaks between words when they run onto the next line. However, on i-Mode phones, text written in the Roman alphabet (such as English and most European languages) gets split when the number of characters on the line is wider than the screen, even if it is in the middle of a word as shown in Figure 5-2.[4]

ACCESSING THE MOST POPULAR I-MODE SITES

The most popular i-Mode sites are by far entertainment-related sites, where you can download character images and ringing tones, play games, read your horoscope, and find dating services. Other popular services include weather information and news-related sites.

So, what about DoCoMo's billing system for i-Mode? Let's take a very brief look.

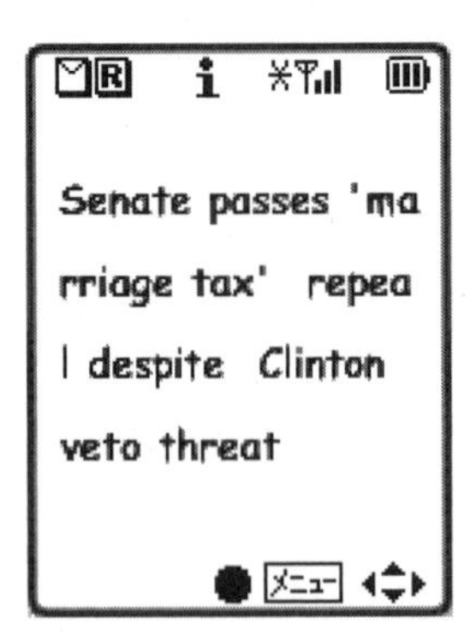

FIGURE 5-2 Text written in the Roman alphabet.

BILLING SYSTEM

Official sites participate in DoCoMo's billing (m-commerce) system. Unofficial sites do not participate in DoCoMo's billing (m-commerce) system. With that in mind, which are the successful i-Mode menu sites?

SUCCESSFUL I-MODE MENU SITES

Examples of very successful i-Mode official sites are Bandai's games and image sites, weather service sites, or for example, a fishing game or sites with surfing information. For example, it will not be trivial work for a foreign company to attempt to build an i-Mode site with similar success as Bandai's sites. At the very least, you will need to build a site that fits Japanese users' tastes, expectations, and needs as well as Bandai's, and also you will need to provide a real value to users, which they can understand and appreciate. You will also need to communicate to the users that you provide this service and the value. Also, you will need to solve all technical issues—but this is the relatively easy part. In addition, you will also need to convince DoCoMo to accept your site as an official menu site.

So, how does one publish an i-Mode-compatible Web site? Let's take a look.

PUBLISHING AN I-MODE-COMPATIBLE WEB SITE

Anybody who publishes an ordinary HTML-based homepage can publish an i-Mode-compatible Web site. As an unspoken convention, many Web sites that offer an i-Mode version in addition to their standard HTML Web site simply create an extra directory called *i* and put their i-Mode-compatible pages there.

For example, the English-language newspaper *Japan Times* has its Web site at **http://www.japantimes.com**, and its i-Mode

Web site at **http://www.japantimes.com/i**. Publishing an official i-Mode site requires the approval of NTT DoCoMo. So, what do you have to do to become an official i-Mode site?

WHAT YOU HAVE TO DO TO BECOME AN OFFICIAL I-MODE SITE

Web sites that can be accessed via i-Mode phones are either official sites, which are registered on the i-Menu, or unofficial sites (also known as *voluntary sites*) without that official sanction. Thus, the advantages of being an official site are that these sites usually get more page views because users can access them from their phone's i-Menu without entering the URL. Also, DoCoMo's billing agency service and unique i-Mode handset IDs are provided only to approved sites.

On the other hand, the disadvantages of being an official site are that search engine services, links to unofficial sites, and community-based content (such as chat and bulletin boards) are not enabled on official i-Mode sites. Nevertheless, once a site is approved, there is no charge for placement on the i-Mode phone menu. All you have to do is to submit your plans to NTT DoCoMo and convince them that your service is unique and useful, that the content is updated regularly, and that you meet the following requirements for English/international information providers as shown in Table 5-1.[5]

Detailed information on how to register an English/international i-Mode site to be listed on the official menu is provided in English on the NTT DoCoMo Web site. Japanese information providers are asked to use the Japanese application page at **http://newip.nttdocomo.co.jp/index.asp**.

So, what picture symbols are allowed on i-Mode sites? Let's look at how you can use picture symbols.

TABLE 5-1 Requirements for English/International Information Providers

Content type	Content planning docs	User (customer) support	Technical support
Japanese/Eng premium	Provide both Eng/Japanese documentation	User support (call center/number) in Japanese; Eng in Japan	Tech support staff to be able to operate in Japanese with (as a minimum) 1 contact in Japan
Eng-only premium	Provide English documentation	User support (call center/number) in Japanese; Eng in Japan	Tech support staff to be able to operate in Japanese with (as a minimum) 1 contact in Japan
Eng/Japanese free	Provide both Eng/Japanese documentation	Considering users, if possible as with the previous, otherwise, e-mail, overseas call number and fax in Japanese/Eng	Tech support staff overseas okay, and in Japanese (1 contact person)
Eng-only free	Provide English documentation	Considering users, if possible as with the previous, otherwise, e-mail, overseas call number, and fax	Tech support staff overseas okay, and in English

PICTURE SYMBOLS ON I-MODE SITES

When Shift-JIS codes are embedded in HTML source code, picture symbols can be described using two-byte hex code as shown in the example in Figure 5-3.[6] However, an editing tool that enables code inputting is needed.

Picture symbols can also be described as text in HTML source code. With this method, six-byte decimal code is used (&#decimal code;). For example, *羚* designates picture symbol number 166 as shown in Figure 5-4.[7] In addition to the

No	Image	S-JIS code		Title
		Hex	Decimal	
1		F89F	63647	Fine
2		F8A0	63648	Cloudy
3		F8A1	63649	Rain
4		F8A2	63650	Snow
5		F8A3	63651	Thunder
6		F8A4	63652	Typhoon
7		F8A5	63653	Fog
8		F8A6	63654	Drizzle
9		F8A7	63655	Aries
10		F8A8	63656	Taurus
11		F8A9	63657	Gemini

FIGURE 5-3 Picture symbols that are allowed on i-Mode sites.

155		F9A4	63908	Mood
156		F9A5	63909	Bad (downward arrow)
157		F9A6	63910	Sleepy (sleep)
158		F9A7	63911	Exclamation mark
159		F9A8	63912	Exclamation & question marks
160		F9A9	63913	Two exclamation marks
161		F9AA	63914	Bump (collision)
162		F9AB	63915	Sweat
163		F9AC	63916	Cold sweat
164		F9AD	63917	Dash (running dash)
165		F9AE	63918	Macron 1
166		F9AF	63919	Macron 2

FIGURE 5-4 Picture symbol number 166.

aforementioned listed picture symbols as shown in Figures 5-3 and 5-4, graphic characters that are part of the standard S-JIS character set can also be used, even in English-language sites.

So, are there any special metatags for i-Mode sites? Let's take a look.

SPECIAL METATAGS FOR I-MODE SITES

To let robots of i-Mode-specific search engines like **iseek.infoseek.co.jp** know that your site's content is i-Mode-compatible, you should include the following metatag inside the head of your i-Mode page:

```
                    <HEAD>
                    <META name="CHTML" content="yes">
                    <META name="description"
content="description in up to 10 double-byte
                    characters">
                    </HEAD>
```

CHARACTER ENCODING ON I-MODE SITES

By default, only Shift-JIS encoding can be used for displaying Japanese characters on i-Mode-compatible Web sites. Some i-Mode handsets might be able to handle other encodings, such as EUC, but if you want to make sure that your content can be seen by all i-Mode users, you should stick to Shift-JIS. For English characters, the standard Western encoding (ISO-8859-1) is allowed.

PC EMULATORS ACCESS I-MODE SITES

A very simple i-Mode emulator in Japanese also works on English Windows systems. If you install the emulator on English Windows, just pressing the default buttons when installing and ignoring the gibberish code should do the trick. Another i-Mode emulator is *i-mode tool 1.0* (also Japanese), which can be downloaded at **http://www.asahi-net.or.jp/~tz2snsmr/**.

No English-language i-Mode emulator is out there yet. However, both the new Japanese and English version of Adobe's home page creation tool GoLive 5.0 incorporates an i-Mode module (see sidebar, "I-Mode Module") that features a preview function for i-Mode-compatible Web sites and an editor for i-Mode-compatible picture symbols (see sidebar, "Adobe

I-MODE MODULE

The i-Mode module gives GoLive 5.0 the ability to visually create NTT DoCoMo mode pages using cHTML. These pages are intended for 501i/502i wireless handsets using the *Emoji* and *i.Form* elements.

Using the i-Mode module within GoLive 5.0 enables the emoji object within GoLive to present real-time renderings of emoji icons, which are only viewable via the 501i and 502i NTT DoCoMo i-Mode wireless handsets. Figure 5-5 shows the GoLive 5.0 wireless authoring module. Pages can be viewed via regular Web browsers, but the emoji entities work only on NTT DoCoMo i-Mode 501i and 502i wireless handsets.[8]

GoLive 5.0 Adds i-Mode and WAP/WML Functionality"). The i-Mode module can be downloaded from the Adobe Web site (**http://www.adobe.com/products/golive/actions/main.html**).

Of course, you can also look at many i-Mode sites from a desktop browser as well (because it's a subset of HTML). Nevertheless, they'll be missing a lot of special characters, and your browser must support Japanese fonts in order to display Japanese sites correctly.

NOTE

Some i-Mode pages will not display at all on a desktop, some will bounce your desktop browser to a different page, and some URLs will be unreachable unless you're accessing them from an i-Mode handset.

USING COOKIES DURING THE DEVELOPMENT OF I-MODE-COMPATIBLE SITES

Finally, can you use cookies during the development of i-Mode-compatible sites? The answer is a resounding no!

FIGURE 5-5 GoLive 5.0 wireless authoring module.

I-Mode does not handle cookies. For session management, you can, however, simulate a cookie-like behavior by appending the desired values (like an ID) to the URL (for example, **http://abc.com/menu.html?id=123&date=0815**). Furthermore, official i-Mode sites can obtain and use the unique handset ID of their visitors for session management.

ADOBE GOLIVE 5.0 ADDS I-MODE AND WAP/WML FUNCTIONALITY

The Adobe's professional homepage creation software for Windows and Mac OS, Adobe GoLive 5.0, incorporates new functionality for the creation of i-Mode and WAP/WML compatible Web sites. The i-Mode module is now available with the shipping version of Adobe GoLive 5.0-J in Japan in October 2000. Additionally, the i-Mode English and WAP/WML modules are now available from **http://www.Adobe.com**.

The GoLive 5.0 i-Mode module supports NTT DoCoMo's compact HTML tagset for the 501i/502i i-Mode handset series (see Appendix F, "Complete Tagset of I-Mode Compatible HTML Tags") and includes the following functions:

- I-Mode-compatible HTML form authoring.
- I-Mode pictures and symbols can be easily dragged and dropped into Web pages and viewed within the GoLive preview window.
- Extensive i-Mode hyperlink management.

With this addition of authoring modules for mobile devices, Adobe is expected to give an important boost to Web designers and developers who are building solutions not only for the desktop, but also for next generation Internet devices. The software list price is 39,800 yen ($371). However, Adobe is promoting the product by offering the first 20,000 copies for 18,000 yen ($168). Adobe is now taking orders for upgrades of GoLive 4.0 and Cyberstudio. The upgrade price is 9,800 yen ($91.50).[9]

CONCLUSION

This chapter discussed the official i-Mode menu sites. In conclusion, how many i-Mode menu sites can you access, and what makes them so special? It all has to do with how the text on an i-Mode menu site is processed, so you can read it on a tiny cell phone screen. I-Mode uses a subtext of HTML called compact HTML to convert the information. Apart from the official content providers, nearly 4,000 sites can be accessed by punching in the proper URL. Many existing Japanese Web site operators are launching i-Mode-enabled sites because it's quite simple to convert an existing Web page to one formatted for i-Mode.

END NOTES

[1]WestCyber Corporation, 1350 Broadway, Rm. 705, New York, NY 10123, 2001.

[2]Ibid.

[3]Ibid.

[4]Ibid.

[5]Ibid.

[6]Ibid.

[7]Ibid.

[8]Adobe Systems Incorporated, 345 Park Avenue, San Jose, California 95110-2704, USA, 2001.

[9]WestCyber Corporation, 1350 Broadway, Rm. 705, New York, NY 10123, 2001.

CHAPTER SIX

DEVELOPING FOR I-MODE

The Internet infrastructure has been developed all over the world, and nowadays a variety of devices are equipped with the Internet-access function, from TV sets to wireless cellular phones. The *HyperText Markup Language* (HTML) is widely accepted and spread as the standard of the *World Wide Web* (WWW) document format. The *compact HTML* (cHTML), discussed in this chapter, defines a subset of HTML for small information appliances such as smart phones, smart communicators, mobile PDAs, and so on. Such a certain level of HTML is strongly required as a guideline from the manufacturers of small information devices, service providers, carriers, and software developers. Because cHTML is completely based on the HTML recommendations, you can use millions of HTML-based content resources, various software tools, and public materials (textbooks, magazines, and Web information).

DEVELOPING CONTENT FOR I-MODE

As you know from discussions in previous chapters, i-Mode uses cHTML, which is a subset of HTML. You need to develop your content in cHTML.

I-Mode also uses special tags in addition to ordinary HTML (a tag that can place a voice call by placing a link, and other extra tags. I-Mode also uses special characters for trains, shinkansen, taxi, kisses, joy, sadness, hot spring baths, and so on, to express emotions in daily life.

COMPACT HTML

As has been previously defined, cHTML stands for *compact HTML* and is the content description language used in i-Mode. CHTML is a subsite of HTML, plus some additional tags, characters, and features. I-Mode also uses special tags (see Appendix F, "Complete Tagset of I-Mode-Compatible HTML Tags"), in addition to ordinary HTML (a tag that can place a voice call by placing a link and other extra tags).

In other words, as previously explained, cHTML is HTML cut down into small chunks, with the omission of some more complex scripts such as tables and frames. This should not discourage the seasoned Web developer vets out there. Almost all of the scripts you currently use are compatible with i-Mode phones, with a few exceptions. A typical phone browser has a maximum width of 16 characters and a height of 6 lines, which leaves no room for some of the more complex features found on a normal browser. Also, all images must be in black-and-white GIF format. Again, you can find a complete list of cHTML compatible tags in Appendix F.

CHTML documents are written exactly as you would write an HTML document. They are single pages, compared to WAP decks, consisting of several cards. The following is a simple example:[1]

```
<html>
<head>
<META http-equiv="Content-Type"
 content="text/html; charset=utf-8">
<META name="CHTML" content="yes">
<META name="description" content="sample cHTML
 document">
```

```
<title>Sample cHTML document</title>
</head>
<body text="#000000">
<center>Links</center>
<hr>
Select an option<br>
<a href="message.chtml" accesskey="1">1</A>
 Messages<BR>
<a href="mail.chtml" accesskey="2">2</A>
 Mail<BR>
<hr>
<center>
<a href="admin@chtmldemo">email: admin@chtmldemo
 </a><br>
</center>
</body>
</html>
```

This should be very familiar to you. The only addition is the use of the accesskey command. This associates the link to the numerical keypad on the phone, making navigation a much easier task.

Because of the backward compatibility of cHTML with newer versions of HTML, supporting i-Mode from an existing Web server is very easy. In this part of the chapter, a description of a simple application for serving news and address book information to users of a community Web site (**http://www.arsdigita.com**), using the *ArsDigita Community System* (ACS) toolkit and platform (**http://www.arsdigita.com/developer**) to author in cHTML will now be presented.

NOTE

The ACS is an open source development platform and suite of prebuilt applications for building database-intensive Web services. ACS consists of a core data model and APIs encompassing content management, e-commerce[2], personalization, collaboration (community and workflow), and administration.

Authoring in CHTML

CHTML is very straightforward to use. It is a small subset of modern HTML, mostly compatible with HTML 1.0 and a small number of HTML 2.0 tags. The sidebar, "CHTML for Small Information Appliances," contains an official list of supported tags and features for an i-Mode terminal. NTT implements some extensions in some of their handsets as well, such as animated GIFs, predefined icons, the Marquee tag, and the autodial Tel: URL prefix. These are documented in NTT DoCoMo (English) supported tags and specs (**http://www.nttdocomo.com/ser2.htm**).

CHTML FOR SMALL INFORMATION APPLIANCES

The cHTML is a well-defined subset of HTML 2.0, HTML 3.2, and HTML 4.0 recommendations, which are designed for small information appliances. HTML defines flexible, portable, and practical document formats for the documents on the Internet. One direction of HTML is to grow toward richer multimedia document format. A new recommendation HTML 4.0 includes new, additional features. For example, *Cascading Style Sheets* (CSSs) give a wider range of document styles. On the other hand, there must be another direction for small information appliances. Small information appliances have several hardware restrictions, such as small memory, low power CPU, small or no secondary storage, small display, mono-color, single character font, and restricted input method (no keyboard or mouse). The browser for cHTML discussed in this chapter can be implemented in such a restricted environment. Once such a subset of HTML is defined, content providers and information appliance manufacturers can rely on this common standard. CHTML definitely contributes to the rapid growth of small information appliance market.

REQUIREMENTS OF SMALL INFORMATION APPLIANCES: SCOPE OF THE PRODUCTS

First, let's describe the scope of target small information appliances. The categories of these devices often are referred to as smart phones, smart communicators, and mobile PDAs. Some hardware restrictions exist for these devices. From the hardware point of view, you should pick up the following main characteristics of the target devices:

- **Small memory** Typical case: 128 to 512KB RAM and 512KB to 1MB ROM.
- **Low power CPU** Typical case: 1 to 10 MIPS class CPU for embedded systems.
- **Small display** Typical case: 50 × 30 dots, 100 × 72 dots, and 150 × 100 dots.
- **Restricted colors** Typical case: mono-color (black and white).
- **Restricted character fonts** Typical case: only single font.
- **Restricted input method** Typical case: several control buttons and number buttons (zero to nine).

Figure 6-1 shows an example of a cellular phone that has the HTML browsing function.[3] A wide variety of content services are potentially possible via wireless networks.[4]

REQUIREMENTS

To realize the WWW browsing function for such small devices, a suitable subset of HTML is necessary. The requirements are derived from the preceding hardware restrictions. Also, these devices should be easy to use from the standpoint of consumer products. The browser software for a subset of HTML should run within the small memory (150 to 200KB for the working data and also 150 to 200 Kbytes for the program code). The minimum requirement for the CPU power should be 1 to 2 MIPS, though it may depend on the CPU power required for network communication processing. Easy navigation is also one of the key features for consumer devices. It means that the users can navigate information with a minimum number of operations. A subset of HTML should satisfy this requirement.

WIRELESS NETWORK

CHTML does not depend on the underlying network protocol. In the typical cases, the transport protocol for cHTML is assumed to be HTTP over TCP/IP. However, current wireless communication

continues

continued

networking for cellular phones is low band and low speed. In this area, the transport protocol should be defined as light protocol for better performance on the physical packet layer. It also seems useful to compress HTML contents so that most of HTML data can be stored within one packet data.

DEFINITION OF CHTML: DESIGN PRINCIPLES

The cHTML is designed to meet the requirements of the small information appliances described in the preceding. It is designed based on the following four principles:

1. *Completely based on the current HTML W3C recommendations.* CHTML is defined as a subset of HTML 2.0, HTML 3.2, and HTML 4.0 specifications. This means that cHTML inherits the flexibility and portability from the standard HTML.
2. *Lite specification.* CHTML has to be implemented with small memory and low power CPU. Frames and tables that require large memory are excluded from cHTML.
3. *Can be viewed on a small mono-color display.* CHTML assumes a small display space of black and white color. However, it does not assume a fixed display space, but it is flexible for the display screen size. CHTML also assumes single character font.
4. *Can be easily operated by the users.* CHTML is defined so that all of the basic operations can be done by a combination of four buttons: Cursor forward, Cursor backward, Select, and Back/Stop (Return to the previous page). The functions that require two-dimensional focus pointing, like *image map* and *table*, are excluded from cHTML.

NOTE

The definition of cHTML is derived straight-forwardly from the preceding principles.

FEATURES OF CHTML

The cHTML is a subset of HTML 2.0, HTML 3.2, and HTML 4.0. The major features are described here, which are excluded from cHTML, as follows:

- JPEG image
- Table
- Image map
- Multiple character fonts and styles
- Background color and image
- Frame
- Style sheet

CHTML includes GIF image support. It should be noted that this subset does not require two-dimensional cursor moving, that is, it can be operated by using only four buttons. Also, well-designed pages for a small display fit the screen space, and the scrolling is not necessary. Actually, the cHTML browser can display the pages like a *deck of cards* by the Handheld Device Markup Language (HDML). Because the memory capacity is the most important issue in implementing the cHTML browser, a buffer limit is recommended for the following functions: Input—The maximum buffer size is 512 bytes, and Select—The maximum buffer size is 4,096 bytes.

Though such a limitation belongs to the implementation issues, the common criteria is useful while developing devices. One recommended implementation for the browser is to support the direct selection of anchors by using number buttons. For example, when five anchors are contained in an HTML page, the third anchor can be selected just by pressing the 3 button.

NOTE

The HTML 4.0 specification includes a new attribute accesskey for the similar purpose of direct key assignment.

continues

continued

DETAIL DEFINITION

The complete list of tags supported by cHTML is described in Appendix G, "Compact HTML Tag List." The comparison with HTML 2.0, HTML 3.2, and HTML 4.0 is also marked. The *document type definition* (DTD) for cHTML is also described in Appendix H, "Compact HTML DTD." This gives the intended interpretation of cHTML elements. The document type is defined as follows:

<!DOCTYPE HTML PUBLIC "-//W3C//DTD Compact HTML 1.0 Draft//EN">

EXAMPLES OF CHTML

Now, let's describe the examples of applications by using cHTML. The following examples show the compact browser for cellular phones. The screen is the space of 7 text lines and is 16 characters wide. The top line is used for displaying the status information. For instance, in the example shown in Figure 6-2, the cursor focus point is expressed as the reverse text.

The example in Figure 6-3 shows the mail sending form-using input tags. The focused form is expressed as solid surrounding lines, and nonfocused forms are expressed as a dotted surrounding line. The cursor point for input characters is expressed as a reverse box.

Finally, the example in Figure 6-4 shows weather and rain information of the day. It uses mono-color GIF image. Nevertheless, practical implementations and experiments show that cHTML is enough-—useful for a small screen of 5 to 10 text lines that is 10 to 20 characters wide.

BENEFITS OF CHTML

The cHTML, an HTML-based approach, guarantees that small information appliances can connect to the open WWW world. CHTML keeps the advantage of HTML features and solves the problems arising from the restrictions of small information appliances.

The cHTML specification can be referred to by the tools like HTML authoring systems. In addition, the client-specific Web services for such small devices can be realized by using user agent attributes. That is, the server can do the content filter for cHTML.

ANOTHER APPROACH

There may be another approach that is not based on HTML standards. The approach of a new language may be accepted in a certain closed service. For example, a language named *Handheld Device Markup Language* (HDML) is proposed for the mobile handheld devices. The goal of HDML is very similar to the one of cHTML. It seems useful for a class of handheld devices. However, the disadvantage of the special language approach can be said that everything such as contents, authoring tools, server software, client software, and textbooks have to be prepared. Especially thinking about a product line from high-end PDAs to low-end cellular phones, the consistent HTML-based approach would make sense.[5]

FIGURE 6-1 Cellular phone with an HTML browsing function.

FIGURE 6-2 CHTML example—simple menu.

FIGURE 6-3 CHTML example—mail send form.

FIGURE 6-4 CHTML example—image contents.

Creating an i-Mode application is very similar to creating a regular HTTP/HTML Web site application. The biggest constraints are the small screen real estate and slow connection speed. Screen size is very small, usually no more than 16 (English) characters by six to eight lines. Phone displays have been black-and-white, but are more commonly now being sold in color (see Figure 6-5).[6]

HTML Page Flow and Forms Ordinary URL hyperlinks can be used in pages and work as expected. Many Web applications make use of HTML forms. Basic HTML forms are supported, including the familiar Input, Select, and Textarea fields. Input fields support types of Text, Checkbox, Radio, Password, Hidden, Submit, and Reset. A guideline of design metrics is available from NTT.

Sample ACS I-Mode Application Next, let's look at how to implement a small ACS application for i-Mode terminals. The application consists of a main menu with two links, one link to a news page, similar to the standard ACS/news module, and the other link to a user directory lookup form (see Figure 6-6).[7]

FIGURE 6-5 The D209I phone weighs 74 g, has a 256 color display, and operates 400 hours (standby receive time) between charging. It will sustain about 120 minutes of talk time on a charge. Data transmission speed is around 9600 bps.

index.html
Main MENU
News
Directory
aD ARSDIGITA

FIGURE 6-6 News page and user directory menu links.

In addition, here is a simple HTML page that is the main menu of the application.[8]

```
<HTML>
<HEAD>
<TITLE>Main MENU</TITLE>
</HEAD>
<BODY>
<FONT COLOR=RED>Main MENU</FONT>
<BR>
<IMG SRC=ad_small.gif ALIGN=RIGHT>
<A HREF=new.tcl ACCESSKEY="1">News</a>
<BR>
<A HREF=addr.tcl ACCESSKEY="2">Directory</a>
</BODY>
</HTML>
```

NOTE

Notice the use of an IMG tag.

GIF images are supported by i-Mode terminals. In fact, color and animated GIF images are now supported. The older recommendations for maximum compatibility were

- Black-and-white 2-bit GIF files are used.
- Only images in GIF 87, 87a, and 89a formats can be used.
- The maximum size of a GIF image should be 94 × 72 dots.[9]

However, phones with color and grayscale displays are rapidly becoming more prevalent. GIF images larger than 94 × 72 pixels will be scaled by the phone, but the entire image is still sent to the phone, so it cannot exceed the 5 KB or 10 KB memory limit, and it is thus wasteful to send images that will be scaled down. Also, the autoscaling can make GIF images hard to read.

The Address Book Function Now, let's look at a way to look up users from the users' table from a quick input form. The form, as shown in Figure 6-7, can be used on an i-Mode terminal.[10]

```
<TITLE>Address Book</TITLE>
Address Book
<FORM ACTION=addr-2.tcl METHOD=POST>
Last Name: <INPUT TYPE=TEXT NAME=last_name><br>
First Name: <INPUT TYPE=TEXT NAME=first_name><br>
Email: <INPUT TYPE=TEXT NAME=email><br>
<INPUT TYPE=SUBMIT NAME=search>
</FORM>
```

You can handle the submitted form with a script (see Code 6-1).[11] The script accesses the users' table in the database and produces an output cHTML page.

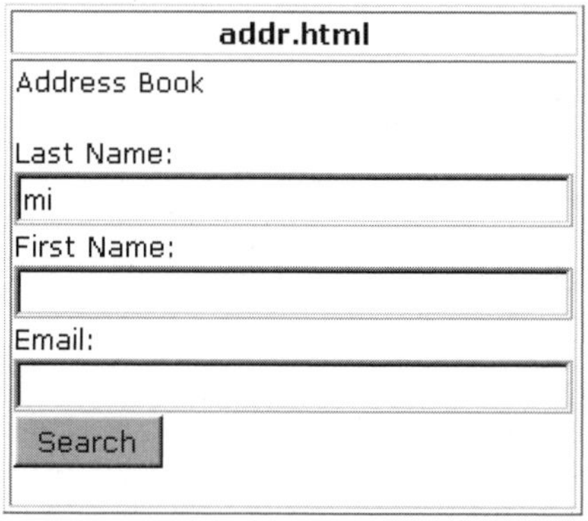

addr.html

Address Book

Last Name:

mi

First Name:

Email:

Search

FIGURE 6-7 Quick input form.

CODE 6-1 Submitted Form Script

```
# addr-2.tcl

imode_basic_auth -uname ""  -pwd [ad_parameter
"GroupPassword" imode]

ad_page_contract {

    Lookup a user name or email in address book

    @param last_name
    @param first_name
    @param email
    @author hqm@arsdigita.com
    @cvs-id spam-add.tcl,v 3.5.2.4 2000/07/21 03:58:01
     ron Exp
} {
    last_name
    first_name
    email
}

#set group_filter_clause " and lower(ug.group_name) =
lower('employees') "
set group_filter_clause ""
```

continues

continued

```
set clauses [list]

if {![empty_string_p last_name]} {
    set pattern "$last_name%"
    lappend clauses " lower(last_name) like
    lower([ns_dbquotevalue $pattern]) "
}

if {![empty_string_p first_name]} {
    set pattern "$first_name%"
    lappend clauses " lower(first_names) like
    lower([ns_dbquotevalue $pattern]) "
}

if {![empty_string_p email]} {
    set pattern "$email%"
    lappend clauses " lower(email) like
    lower([ns_dbquotevalue $pattern]) "
}

set page_content ""

if {[llength $clauses] == 0} {
    append page_content "You must enter a search term"
} else {
    set query "select distinct users.user_id as uuid,
        users.last_name,
        users.first_names,
        users.email,
        uc.home_phone,
        uc.work_phone,
        uc.cell_phone,
        uc.pager,
        uc.fax
   from users, users_contact uc, user_groups ug,
   user_group_map ugm
  where users.user_id = ugm.user_id
        and ug.group_id = ugm.group_id
        $group_filter_clause
        and users.user_id = uc.user_id(+) and [join
        $clauses " AND "]
  order by last_name
      "
    set entries ""
```

```
    set i 1
    db_foreach matching_items $query {
        append entries "<br>$i. <a href=\"one-
        user.tcl?[export_url_vars uuid]\"
        accesskey=\"$i\">$last_name, $first_names</a>
        <a href=\"mailto:$email\">$email</a>"
        incr i
        # limit results to size of phone page memory
        if {[string length $entries] > [ad_parameter
        MaxPageSize imode 4500]} {
            break
        }
    }

    if {[empty_string_p $entries]} {
        append page_content "No matches"
    } else {
        append page_content $entries
    }
}


ns_return 200 "text/html; charset=shift_jis" "
<title>iMode Address Book</title>
$page_content
"
```

The script could be modified to restrict the search to only users in a specific group, such as the employees of a company, or only members who share the same group as the user as shown in Figure 6-8.[12]. The output of the search script looks like the following:[13]

```
<title>i-Mode Address Book</title>
<br>
1. <a href="one-user.tcl?uuid=41"
accesskey="1">Miller, Mark</a>
<a
href="mailto:mark@yahoo.com">mark@yahoo.com</a>
<br>
2. <a href="one-user.tcl?uuid=22"
accesskey="2">Minsky, Henry</a>
<a href="mailto:hqm@ai.mit.edu"hqm@ai.mit.edu</a>
```

addr-2.tcl

1. Miller, Mark
mark@yahoo.com
2. Minsky, Henry
hqm@ai.mit.edu

FIGURE 6-8 Users in a specific group.

NOTE

The accesskey attribute of a hyperlink provides one-key access to select, and follows the URL from the phone's numeric keypad.

You can also define a suggested maximum page size configuration parameter, so that pages that can potentially generate a lot of data, such as a search results page, can have an idea of when to truncate their results. Sending more data in a page than the phone can support results in an error message and possibly an unreadable page. Even if a phone supports a 10-KB cache size, it is still desirable to limit the amount of text on a single page as much as possible because scrolling through 10 KB on a tiny, slow browser is not much fun.

Display One User You can display a single user's (see Figure 6-9)[14] information using a script (see Code 6-2)[15] that produces output like this:[16]

```
<title>Minsky, Henry</title>
Minsky, Henry
<br>
<a href="mailto:hqm@ai.mit.edu">hqm@ai.mit.edu</a>
<br>
1. home phone: <a href=tel:81357173259
accesskey=1>+813 5717 3259</a>
<br>
2. work phone: <a href=tel: accesskey=2></a><br>3.
cell phone: <a href=tel:accesskey=3></a><br>pager:
<br>fax:
<br>
```

one-user.tcl

Minsky, Henry
hqm@ai.mit.edu
1. home phone:
+813 5717 3259

2. work phone:
3. cell phone:
pager:
fax:

FIGURE 6-9 Displaying a single user's information.

CODE 6-2 A Single User's Information Using a Script

```
# one-user.tcl

ad_page_contract {

    Lookup a user's contact info

    @param  uuid user id
    @author hqm@arsdigita.com
    @cvs-id spam-add.tcl,v 3.5.2.4 2000/07/21 03:58:01
    ron Exp
} {
    uuid:integer
}

# args: uuid

imode_basic_auth -uname ""  -pwd [ad_parameter
"GroupPassword" imode]

proc imode_phone {str} {
    # replace spaces with -
    regsub -all  {[^0-9]} $str "" str
    return $str
}
```

continues

continued

```
set query "select users.user_id as uuid,
        users.last_name,
        users.first_names,
        users.email,
        uc.home_phone,
        uc.work_phone,
        uc.cell_phone,
        uc.pager,
        uc.fax
   from users , users_contact uc
  where users.user_id = uc.user_id(+)
        and users.user_id = :uuid

    "
db_1row user_stuff $query

append page_content "$last_name, $first_names<br>"
append page_content "<a
href=\"mailto:$email\">$email</a><br>"
append page_content "home_phone: <a
href=tel:[imode_phone $home_phone]
accesskey=1>$home_phone</a><br>"
append page_content "work_phone: <a
href=tel:[imode_phone $work_phone]
accesskey=2>$work_phone</a><br>"
append page_content "cell_phone: [imode_phone
$cell_phone]<br>"
append page_content "pager: [imode_phone $pager]<br>"
append page_content "fax: $fax<br>"


ns_return 200 "text/html; charset=shift_jis" "
<title>iMode Address Book</title>
$page_content
"
```

The *tel:* protocol prefix for a URL tells the phone (browser) that the link is a phone number that can be dialed. If the user selects the link, a phone call is placed.

THE WHAT'S NEW PAGE The News page (see Figure 6-10)[17] links to a script (see Code 6-3)[18] that calls the **http://www.arsdigita.com/doc/news** ACS news module and produces a compact listing of recent events:[19]

```
<b>News</b>
<br>
Jul 27, 2000:
<a href="one-news-
item?news_item_id=100020">ArsDigita provides new
services</a>
<br>
Jul 26, 2000:
<a href="one-news-
item?news_item_id=100000<\#34>>i-Mode now on
imode.arsdigita.com</a>
```

new.tcl

News
Jul 27, 2000:
ArsDigita provides new services
Jul 26, 2000:
i-Mode now on imode.arsdigita.com

FIGURE 6-10 News page links.

CODE 6-3 The News Page Links to a Script

```
# imode/new.tcl

ad_page_contract {

    Gets the news from the news module

    @author hqm@arsdigita.com
```

continues

continued

```
    @cvs-id spam-add.tcl,v 3.5.2.4 2000/07/21 03:58:01
    ron Exp
} {
}

set since_when [db_string since_when "select sysdate -
30 from dual"]

# modify this to get user_id from query arg
set user_id [ad_get_user_id]

append page_content "
<b>News</b>
"
set users_table "users"

# What is the maximum number of rows to return?
# the how_many proc argument has precedence over the
value in the ini file
set max_stories_to_display [ad_parameter
DefaultNumberOfStoriesToDisplay imode 10]

set newsgroup_clause "(newsgroup_id = [join
[news_newsgroup_id_list $user_id 0] " or newsgroup_id =
"])"


set query "
select news.title, news.news_item_id,
news.approval_state,
       expired_p(news.expiration_date) as expired_p,
       to_char(news.release_date,'Mon DD, YYYY') as
       release_date_pretty
  from news_items news, $users_table ut
 where creation_date > '$since_when'
       and news.creation_user = ut.user_id
       and $newsgroup_clause
 order by creation_date desc"

set result_items ""
set counter 0
```

```
db_foreach news $query  {
    append result_items "<br>$release_date_pretty: <a
    href=\"one-news-
    item?news_item_id=$news_item_id\">$title</a>"
    incr counter
    if { $max_stories_to_display > 0 && $counter >=
    $max_stories_to_display } {
        break
    }
}

append page_content $result_items

doc_return 200 "text/html; charset=shift_jis"
$page_content
```

A single news item is formatted in as compact a manner as possible, using the one-news-item.adp (see Figure 6-11)[20] script (see Code 6-4).[21]

So, can you send someone an i-Mode emulator? Let's see.

FIGURE 6-11 A single news item is formatted.

CODE 6-4 One-News-Item.adp Script.

```
#
# /www/imode/one-news-item.tcl

ad_page_contract {

    News item page

    @author hqm@arsdigita.com
    @cvs-id spam-add.tcl,v 3.5.2.4 2000/07/21 03:58:01
    ron Exp
} {
   news_item_id
}

set nrows [db_0or1row news_item "
select title, body, html_p, n.approval_state,
release_date,
        expiration_date, creation_user, creation_date,
        first_names, last_name
from news_items n, users u
where news_item_id = :news_item_id
and u.user_id = n.creation_user"]

if { $nrows == 0 } {
    ad_return_error "Can't find news item" "Can't find
    news item $news_item_id"
    return
}

append page_content "
<title>$title</title>
<body>
<b>$title</b>
<p>
[util_maybe_convert_to_html $body $html_p]
<p>
Contributed by $first_names $last_name

"

doc_return 200 "text/html; charset=shift_jis"
$page_content
```

I-MODE EMULATOR

Japanese language i-Mode simulators do exist, but no one knows of any English language i-Mode simulators, except for an *i-Mode demonstrator* on the French Web pages of N-I-I—DoCoMo France. You can download a Japanese language i-Mode simulator. However, these simulators are not sufficient for serious commercial developments.

Some Web-based i-Mode simulators in Japanese language are available, but you cannot download them. The reason is that 99.9 percent of i-Mode users are Japanese, almost all the development is done in Japan, and in Japan, it is very easy to use an i-Mode phone for development. Also, there are many different handsets, and you cannot really get a full feeling of what users experience, unless you use a real i-Mode handset.

Commercial i-Mode developers in Japan, in addition to real i-Mode handsets, do use simulators; however, these are not for general sale. NTT-DoCoMo's resources for i-Mode/cHTML developers can be found at the following sites/URLs:

- DoCoMo's description of cHTML, explanation of HTML 1.0, and HTML 2.0 compatible tags (**http://www.nttdocomo.com/i/tag/**)
- Outline of i-Mode compatible cHTML (**http://www.nttdocomo.com/i/tag/imodetag.html**).
- List of i-Mode compatible HTML 2.0 tags (**http://www.nttdocomo.com/i/tag/lineup.html**)
- A list of DoCoMo's special symbols (**http://www.nttdocomo.com/i/tag/emoji/index.html**)
- DoCoMo's resources for developing for JAVA on i-Mode (**http://www.nttdocomo.co.jp/i/java/index.html**)

WARNING

URLs may change without notice.

USING CHARACTER SETS FOR I-MODE

Almost all users of i-Mode are Japanese. Therefore, you need to write your content in the Japanese language using Japanese characters. In other words, it is important to use Japanese symbols in your cHTML pages. In addition, you can also get a set of 166 hexadecimal symbols to enable the developer to display some common *emoticons*, or symbols representing trains, shinkansen, taxis, kisses, joy, sadness, hot spring baths, and so on, to express emotions.

NOTE

I-Mode uses some special symbols and characters (for kisses, hot baths, Shinkansen trains, and so on), which are not part of the normal set of Japanese characters. On a Japanese-enabled Web browser, these special characters will shoot up as a question mark. I-Mode uses Shift-JTS encoding.

I-Mode phones also support the Shift-JIS character set, which fortunately, is backward-compatible with 7-bit ASCII. Most standard English text will display without difficulty on an i-Mode terminal.

FORMS AND E-MAIL

CHTML enables the use of simple forms and can be used to send and receive e-mail. All of the form tags are the same as found in traditional HTML. These are two of the most important features that can bring the average consumer into the wireless world. I-Mode phones enable the user to send and receive any e-mail from anywhere in the world, but are also limited in size. E-mails must be less than 500 Roman characters or 250 double-byte Kanji characters and cannot contain any attachments.

NOTE

Emoji means picture characters. These are special characters for trains, shinkansen, taxi, kisses, joy, sadness, hot spring baths, and so on, to express emotions and other things, some of which are specific to Japan. There are 166 emoji.

If you would like to send an image in the e-mail, simply add the address where the image can be found into the body. The default e-mail address of i-Mode users in Japan is **090xxxxxxxx@docomo.ne.jp**, where 090xxxxxxxx is the telephone number. E-mails are the most popular usage of i-Mode phone users around the world, so do not be afraid to add them to your pages.

NOTE

You cannot send images by e-mail. However, you can e-mail the URL of a GIF image, which when clicked by the recipient of the e-mail, will display the image on the handset, and the image can be downloaded into the handset.

EXCLUDED FEATURES

Due to file size constraints, limited screen sizes, and overall lack of user friendliness, some features found in HTML cannot be used under any circumstances. These include

- JPEG images
- Tables
- Image maps
- Multiple character fonts and styles

- Background color and images
- Frames
- Style sheets[22]

TESTING

Probably one of the biggest advantages to developing in cHTML is the ability to view and test your sites in any Web browser, with results only slightly differing from the actual devices. This supercedes many of the problems associated with WAP development, especially for Mac users. No English language simulators are available yet, but this should change by the end of 2002.

POINTS TO REMEMBER

Finally, here's a list of some important issues to keep in mind when developing i-Mode format pages:

- Images must be in GIF format.
- Only Shift-JIS text coding is supported.
- You can only scroll vertically.
- Relative paths and absolute paths can be used in URLs.
- The total file size for any page, including images, must be less than 5K (2K is recommended).

CONCLUSION

This chapter discussed developing for i-Mode. In conclusion, cHTML was recommended for small information appliances. This contributes to the WWW community, especially for wireless and mobile small devices. CHTML can be referred as a recommended guideline for HTML services and for HTML browsing software in this area.

The Internet world is growing very fast. Several levels will likely be required for cHTML in the near future. It is important to evolve the functionality and extend the specification flexibly, according to the requirements from the growing market.

In the very near future, the worldwide standard for wireless digital phone networking protocols will be established. The bandwidth of the network is expected to be wide enough for motion picture (video) communications. CHTML should include such advanced features in this next generation of wireless network.

Finally, developing for any mobile device requires strict attention to detail. Any inconsistencies in your syntax will result in a nasty error on the user's phone. The mobile Internet is an emerging technology that will undoubtedly continue to change throughout the years, so it is important to have an understanding of all used languages, including cHTML.

END NOTES

[1]Jay Staton, "Transitioning from HTML to CHTML Development," AnywhereYouGo.com, 3000 Waterview Parkway, B2E14, Richardson, TX 75080, 2001.

[2]John R. Vacca, *Electronic Commerce: Online Ordering and Digital Money with CD-ROM*, Charles River Media, 2001.

[3]Tomihisa Kamoda, "Compact HTML for Small Information Appliances," AnywhereYouGo.com, 3000 Waterview Parkway, B2E14, Richardson, TX 75080, 2001.

[4]John R. Vacca, *Wireless Broadband Networks Handbook: 3G, LMDS, and Wireless Internet*, McGraw-Hill, New York, 2001.

[5]AnywhereYouGo.com, 3000 Waterview Parkway, B2E14, Richardson, TX 75080, 2001.

[6]ArsDigita Corporation, 80 Prospect Street, Cambridge, Massachusetts 02139,2001.

[7]Ibid.

[8]Ibid.

[9]Ibid.

[10]Ibid.

[11]Ibid.

[12]Ibid.

[13]Ibid.

[14]Ibid.

[15]Ibid.

[16]Ibid.

[17]Ibid.

[18]Ibid.

[19]Ibid.

[20]Ibid.

[21]Ibid.

[22]Jay Staton, "Transitioning from HTML to CHTML Development," AnywhereYouGo.com, 3000 Waterview Parkway, B2E14, Richardson, TX 75080, 2001.

CHAPTER SEVEN

I-MODE FOR BUSINESS

I-Mode cell phones have proven that companies can make money on wireless Internet access[1], and with their explosive popularity, the phones could one day rival the number of PCs on the Internet. NTT DoCoMo's i-Mode cellular service (which lets consumers receive e-mail and browse the Web) is gaining 60,000 new subscribers a day, or 1.8 million a month.

The i-Mode has the business potential to catch up with the PC Internet market. The i-Mode is not only useful for the consumer market but as a productivity tool.

NOTE

The number of PC users, of course, dwarfs the number of I-mode subscribers.

The i-Mode growth rate, however, is far more rapid. The service only started in February 1999, and since then, has garnered over 27 million subscribers as of this writing, all in Japan. International expansion is under way, and the Japanese wireless carrier has recently invested significantly in AT&T Wireless.

Mobile Internet access is expected to be a huge market, with thousands of companies targeting the emerging sector. But with the exception of i-Mode in Japan, U.S. wireless Net access has been something of a bust so far.

NTT DoCoMo is also set to make a monster offering on the Tokyo stock exchange. According to Reuters, the wireless carrier aims to raise around $7.07 billion (889 billion yen) with an offering of 500,000 new shares priced at a discounted 2.066 million yen each—the biggest offering by a listed Japanese company.

DoCoMo indicated the funds will be used for the company's expansion drive into foreign markets. The company will enable individual investors to buy 330,000 shares. Japanese institutional investors can buy up to 50,000 shares, while investors in the United States and elsewhere are allocated 80,000 each.

Just as important as the growth in subscribers, i-Mode is attracting developers, e-commerce companies,[2] and others to provide offerings for the service. Approximately 887 developers write applications for i-Mode, and 519 different search engines have become i-Mode friendly. More than 48,000 sites now provide i-Mode content, and subscribers churn through an average of 20 page views per site visit.

Content services have also sprouted up around the phones. Approximately 200 Web sites a day tweak their sites so i-Mode subscribers can download information. News services have launched subscription services that deliver breaking headlines to customers for $1 to $3 a month. So far, more than 200,000 subscribers exist. In addition, corporations have begun to adapt the phones for remote data access, a concept Palm is pushing.

The i-Mode will become the simplest remote access (device) in the bunch. Nonetheless, this is a huge market, and no one knows yet who will become a winner.

The next phase in development will occur in early 2002, when the company unveils its first *third-generation* (3G) wireless service, which will deliver data between 64 and 384 Kbps. At these rates, it will become possible to deliver songs or video over wireless networks. Restaurant location programs will also be able to deliver 3D maps of the restaurant that describe the ambiance.

The cell phone will eventually become a key to selling products and include pictures and text. Along with increased bandwidth, the phones themselves will get brawnier. Faster processors, more memory, and screens with finer resolution will all be part of the picture.

The push on i-Mode, which was conceived in late 1996, is a way to stave off slowing subscription rates. Only about 90 million Japanese citizens will own cell phones (70 million already do). I-Mode enables NTT to place less emphasis on fighting over new customers and more emphasis on selling customers more expensive services.

Competition (in the standard cell phone market) will become much more severe. Income will be saturated in the near future.

HOW IS I-MODE USED FOR BUSINESS?

You can sell content on i-Mode you can sell services via i-Mode: airlines sell air tickets via i-Mode, media companies sell cartoon images via i-Mode, securities houses sell shares and investments via i-Mode, and Japanese government lottery tickets are sold via i-Mode, just to name a few examples. You can have a virtual private network on i-Mode, and many other business applications are for i-Mode.

IMPACTS TO INFORMATION TECHNOLOGY

I-Mode proved that the wireless Internet connection is so easy. Although a PC takes more than 60 seconds to boot up Windows and connect to the Internet, i-Mode needs less than 10 seconds to go there. DoCoMo's 27 million subscribers, plus the competitors' 20 million, are 47 million (see sidebar, "DoCoMo's I-Mode Soars Past 27 Million Subscribers"). This is almost equivalent to PC Internet users in Japan. Before i-Mode, Japanese Internet penetration was 13.4 percent (1998), whereas the United States was 37.0 percent. The latest survey predicts that penetration will be 67.6 percent and 59.6 percent in 2002, respectively.

DOCOMO'S I-MODE SOARS PAST 27 MILLION SUBSCRIBERS

The i-Mode wireless Internet service of Japan's NTT DoCoMo has topped 27 million subscribers (as of this writing)—more than five times the 5 million signed on to the service in March 2000. I-Mode, a proprietary always-on mobile Internet service supported by thousands of compatible Web sites, is currently the most successful wireless Internet service in the world.

As previously explained, it was only launched in February 1999 and is currently adding a 1.8 million new subscribers each month. It has become Japan's largest Internet service provider.

Content and commerce partners use the i-Mode portal to provide mobile online services to i-Mode users, including wireless banking, news, games, and more. Today, 602 companies are providing services through i-Mode, with about 2,000 official i-Mode Web sites and a further 29,800 independent i-Mode compatible sites.

The Japanese carrier (DoCoMo) has been looking at ways to expand the service, including through a link-up with the PlayStation video games console (see sidebar, "PlayStation Connects to Net Via I-Mode"). As previously discussed, i-Mode is based on a packet-data transmission system, with subscribers charged according to the volume of data they transmit, not the time they are on the line. The mobile terminal is constantly in communication with the i-Mode server on DoCoMo's network. This server, in turn, has links to information providers via the Internet or leased lines.

Web sites targeted at i-Mode users can be created in standard HTML, with some small adjustments for compatibility. DoCoMo checks data pushed at i-Mode users for billing purposes and to ensure it has been requested by the mobile user.[3]

I-Mode is also changing the Net business. NTT DoCoMo has become the biggest Internet provider in Japan. AOL took 15 years to have 24 million subscribers; however, NTT DoCoMo obtained 27 million customers within just 30 *months*. If more people access from cell phones than from PC, DoCoMo's portal site would be more attractive than Yahoo! would.

PLAYSTATION CONNECTS TO NET VIA I-MODE

Sony PlayStations sold in Japan are now connecting to the Internet via the i-Mode service. The two companies have agreed to jointly develop a new service that combines the game console with NTT DoCoMo's i-Mode service, which is popular in Japan. The companies hope the collaboration will inspire software developers to create new games and services that take advantage of the integration. The companies have agreed to jointly combine standard i-Mode service, Java-enabled i-Mode service, and faster-bandwidth W-CDMA service with PlayStation.[4]

Market potential for i-Mode is huge. Japan's market potential is estimated at 470 million units, which is three times as many as the entire population by 2010. Japan's wireless market generates $60 billion in revenue. An industry association estimated the revenues in 2003 would be $200 billion.

LANGUAGES THAT I-MODE COMES WITH

As previously discussed, you need to know the Japanese language to do business with i-Mode. If you do not understand Japanese and do not read and speak Japanese, you can still use i-Mode to address foreigners in Japan who do not speak Japanese—and, your preference to read and do business in English. You will find, however, that quite a few foreigners working in Japan actually use the Japanese sections of i-Mode. At the moment, most business applications of i-Mode are in the Japanese language.

English Language Section of the Official I-Mode Menu

The official i-Mode menu contains an English language section. It was prepared for the G8 meeting in Okinawa in summer 2000, but the list of content is different and much, much smaller than the Japanese language section.

English-Language Instructions for Using My I-Mode

Some i-Mode phones come with a short English summary at the end of their instruction manual, but most provide only a basic introduction. The N209i (**http://www.nttdocomo.co.jp/i/lineup/n209i/n209i.html**) as shown in Figure 7-1, has a bilingual menu, except for the list of Japanese language sites and the *options* setting page.[5] It also comes optionally with an English-language manual. On the other hand, the NM502i i-Mode phone from Nokia (**http://www.nttdocomo.co.jp/i/lineup/nm502i/nm502i.html**) as shown in Figure 7-2, is fully bilingual and also comes with an English-language manual.[6]

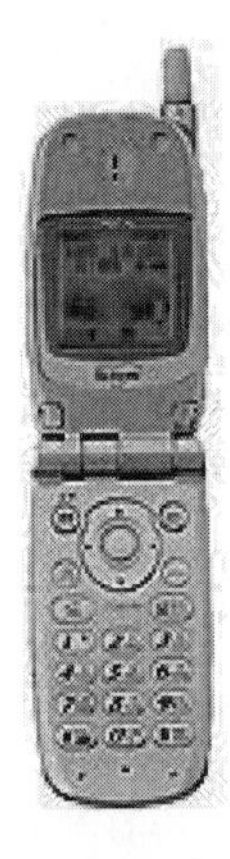

FIGURE 7-1 The N209i i-Mode phone.

FIGURE 7-2 The NM502i i-Mode phone.

Finally, let's take a very brief look at an i-Mode for business case study. This case study ties the preceding discussion together, so that it makes perfect sense to use i-Mode technology for business applications.

I-MODE BUSINESS CASE STUDY: DAI-NIPPON PRINTING CO.

Dai-Nippon Printing (DNP)[7], one of the world's biggest printing companies, launched mobile sales support system in October 1999. The system consists of Dynagalaxy (Integrated database) and Value mail (DoCoMo's service package to connect mobile phones to databases). Dynagalaxy includes all data for DNPs in forms of text, picture, video, and sound. It can also output these data as printed materials, CD-ROMs, Web pages, and presentation files. Value mail connects smart phones to the database in 10 seconds without creating new client servers or setting up a new firewall.

The new system enabled real time publishing. Before the new system, a sales representative had to wait long intervals between each process. For example, a writer sends his/her draft to his/her sales rep. She/he brings it to a proofreader and sends it back to the writer. The writer authorizes the revised draft and sends it back to her/him. She/he then sends the draft to a designer and sends it back to the writer for his/her admission. After the admission, the writer sends it back to her/him and the rep transfers it to a printing factory.

Now, some of these transactions are dealt with the new system. The draft was e-mailed to the database. The sales rep knows the arrival with a smart phone and she/he asks the proofreaders to check the draft. The proofreader picks up the file from the database and e-mails a revised version file to the database. The sales rep e-mails the new version file to the writer and the designer has already started designing process. The new system saves the waiting time between the processes, and the sales rep can monitor every process with her/his smart phone.

DNP is extending this idea to other areas, such as procurement procedures and information technology, inside the company. For example, the sales reps would settle their transportation fees with the system, request meeting rooms or cars, download daily, weekly, monthly reports, and have a brain storming with their smart phones.

This case study demonstrates i-Mode's four impacts to information technology: accessibility, personalization, information sharing, and authority decentralization. The sales rep can access everybody in charge of the project from anywhere in any time. The project is not carried out by company-to-company or division-to-division base. Rather, each involved individual worked as a project team. I-Mode destroys the difference between insiders and outsiders. They can save their time by sharing information in the same platform.

This is a completely different paradigm from the old economy, which adds values by shutting down a company's information from the outside. In this paradigm, middle management turns out to be another headache, because sales reps must wait for its permission while they can forward their draft to anywhere in the world with a touch of a button. Finally, although legacy information technology is aimed to support the decision making of headquarters, the new information technology in an i-Mode era should support the decision making of sales representatives.

CONCLUSION

This chapter discussed the use of i-Mode for business. In conclusion, despite the success of the i-Mode mobile service for business, Internet use in Japan is abnormally low for a developed country. Recently, NTT launched L-mode, a proprietary data service that uses an HTML-like interface and a screen phone with e-mail capability and telephony tied in. However, L-mode is primitive and has hit a brick wall of opposition from competitive operators.

Japan's NTT is set to launch a fixed-line version of its successful i-Mode service. NTT calls the service L-mode and plans to achieve with it what the government has been unable to do with the Internet, get every household in Japan networked to a data service. However, the government and NTT's competitors are deeply concerned that L-mode may help reinforce NTT's virtual monopoly over the local loop and restrict competition in the provision of online services.

Nearly 20 years after France launched its Minitel experiment, NTT's new L-mode service looks remarkably similar in concept. Minitel was introduced as a directory information service before anyone, outside of the Pentagon, had heard of the Internet. Minitel became hugely successful, not least as France Telecom gave away or heavily subsidized the proprietary terminals to use it. The only drawback was that when most of the Western world started logging onto their favorite Web sites, the French logged onto the Internet uptake, because they already had some of the most useful information and services available through Minitel.

That is now well in the past and the Internet has become as ubiquitous in France as elsewhere in the Western world. Moreover, the Minitel experience can be seen as having instilled a concept of networked information in the French much earlier and more broadly than elsewhere.

But, the motivation for the introduction of NTT's L-mode is substantially different from 20 years ago in France. Japan's planners have identified several concerns in the growth of online data services. First, despite the hype over the wireless Internet and i-Mode, the Internet has not become an everyday tool for the Japanese as it has elsewhere. In fact, Japan still languishes near the bottom of the OECD league table of countries when it comes to Internet use. A second concern is seen as balancing a need to foster a competitive telecom environment with the interests of Japan, and the world's largest company by market capitalization (NTT).

NOTE

Several practical reasons account for the poor uptake of the Internet in Japan: the Internet is still fundamentally an English-language application; Japanese houses are small, and there is simply not the room to dedicate space for a home computer; work and social habits are very different in Japan than from elsewhere; and there is still a need to disassociate home environment totally from the workplace environment.

However, major faults can also be found in each of these reasons. For example, Japan's neighbor, South Korea, has similar language issues, domestic space confines, and some social practices. However, South Korea has what some claim to be the most connected society on earth. Why the difference?

Part of the problem is that the Internet is a revolutionary tool and innate conservatism is in Japan that resists social change. One example of this is how the Japanese government legalized the contraceptive pill in 1999, nearly 30 years after Western countries had accepted it as an everyday health measure. The pill was only legalized when Japanese women were given a forum to find information and question relevant authorities.

Social conventions, including the structure of the Japanese language, had excluded Japanese women from a range of information. Of course, change is inevitable and the Internet is already causing social upheaval in Japan. However, it is not changing as quickly as may be expected.

Japan still has Asia's smallest proportion of female Internet users to male users, even though wireless Internet access through i-Mode has substantially increased the overall level of Internet use in Japan. The demographic in Japan with the fastest Internet uptake are young women.

NTT has picked up on this in its marketing material for L-mode, arguing that the L stands for lady, local, or living. Lady, because NTT hopes the service will specifically appeal to Japanese women who do not have regular access to the

Internet. Living, because the service will enhance everyday living standards by giving all Japanese households access to online banking, ticket reservations, news, and e-mail services. Local, because the service will offer community-based content, such as information on sales at local stores, events at local schools, and local government services.

L-mode will deliver content based on cHTML. Subscribers have to buy an L-mode handset equipped with a liquid crystal display. NEC, Matsushita Electric Industrial, and Matsushita Communication Industrial have all released L-mode handsets. Sharp and others are due to make new products available soon in time for both the service's launch and employees' bonuses.

The L-mode handsets have functions for browsing text-based Web pages and for sending and receiving e-mails. The user dials into the nearest L-mode access point. From there, the call is routed via the *pstn* to either of the two L-mode gateways set up by NTT's two regional carriers, NTT East and NTT West. From the gateway, the call is connected to the Internet.

An e-mail server is installed at each of the gateways with e-mail addresses consisting of a user's telephone number followed by @pipopa.ne.jp. Messages of up to 2,000 characters can be sent and received.

Content providers can either make their content available over the Internet or connect their own server via a leased line to the gateways. NTT says a range of free and paid for services will be provided, such as weather forecasts, ticketing, and online trading. The two regional arms of NTT have set the basic usage fee for the L-mode service at around $4 a month, although separate connection charges will also apply. NTT predicts it will have 2.6 to 3 million L-mode subscribers by the end of 2002.

The timing of the launch of L-mode is controversial and has touched on issues that are at the heart of debate over the direction of competition in Japan's telecom sector. When NTT applied in February 2001 for permission to launch the service, the application was rejected on the grounds that the business plans would conflict with a law limiting NTT to intra-prefectural service.

NTT resubmitted the modified applications, under which other telecom carriers will manage the servers for transmissions outside prefectural borders and set rates for such services. The Japanese government accepted the modified business plan and set down guidelines for NTT to show menu lists of other Internet service providers on the first screen and to make the process of selecting servers fair and transparent to try to guarantee fair competition.

However, NTT's competitors complain that even with these modifications, L-mode will effectively deliver NTT a monopoly over domestic Internet access in Japan. KDDI, for example, argues that as NTT owns the copper's last mile (and few Japanese households have Internet connections), then NTT is ensuring that it controls domestic Internet access. NTT's response has been to present a new plan that will enable competition in its fiber to home market.

The plan will enable businesses to lease fiber-optic connections into buildings at a rate of around $40 per circuit. This would enable telecom businesses, including Internet-related venture firms, the ability to offer households a high-speed Internet connection without having to set up their own networks. NTT indicates that the move is in response to the government's e-Japan strategy, aimed at making Japan a far more wired nation by providing an ultra high-speed Internet connection service accessible to 20 million households by 2006.

Yet, NTT indicates it is not obliged to make its fiber-optic cable networks[8] accessible to other telecom operators. It is currently connecting fiber-optic cables from local telephone exchanges to access points, which have been installed at a ratio of 1 for every 400 households nationwide.

The companies have offered 60-percent coverage in major cities and prefectural capitals and 40 percent elsewhere. In December 2000, fiber-optic cables were extended to households from the access points.

The pieces are all in place for an almighty battle to win Japan's huge consumer market for domestic Internet use in the future. Yet, although NTT still has significant monopoly advantages, L-mode is so primitive, that even if it is not a disaster from day one, its shelf life will be strictly limited. More important is what sort of access competitors will have in providing tomorrow's broadband services.

END NOTES

[1]John R. Vacca, *Wireless Broadband Networks Handbook: 3G, LMDS, and Wireless Internet*, McGraw-Hill, New York, 2001.

[2]John R. Vacca, *Electronic Commerce: Online Ordering and Digital Money with CD-ROM*, Charles River Media, 2001.

[3]ComputerUser.com, Inc., 1250 Ninth Street, Berkeley, CA 94710, 2001.

[4]INT Media Group, Inc., 23 Old Kings Highway South, Darien, CT 06820, 2001.

[5]5 WestCyber Corporation, 1350 Broadway, Rm. 705, New York, NY 10123, 2001.

[6]Ibid.

[7]Dai Nippon Printing Co.,Ltd, 1-1, Ichigaya-Kagacho 1-Chome, Shinjuku-ku, Tokyo 162-8001, 2001.

[8]John R. Vacca, *The Cabling Handbook* (2nd Edition), Prentice Hall, Upper Saddle River, N.J., 2001.

PART III

INSTALLING AND DEPLOYING I-MODE

CHAPTER EIGHT

I-MODE VERSUS WIRELESS APPLICATION PROTOCOL (WAP)

With common knowledge dictating that the United States is far behind Japan in terms of mobile Internet use, it's no wonder some insiders might feel NTT DoCoMo's popular i-Mode technology will beat the stuffing out of the *Wireless Application Protocol* (WAP). This is not to say i-Mode versus WAP is a tug-of-war between America and Japan (WAP's provenance is actually of international flavor as Unwired Planet, Motorola, Nokia, and Ericsson are all founding fathers), but rather a matter of which technology is better. Analysts feel WAP, a specification that theoretically enables users to access information via mobile phones, pagers, and other various and sundry handhelds, is seriously lagging behind the proven i-Mode, which is a complete brand of wireless Internet access.

Boasting over 28 million subscribers and counting, i-Mode users enjoy sending and receiving e-mail, exchanging photographs, shopping, banking, downloading personalized ringing melodies for their phones, and navigating thousands of specially formatted Web sites. Though it won't start your coffee machine for you in the morning yet, or tell you the latest sales at the J.Crew a couple blocks away, the sky is the limit in the years to come for i-Mode, analysts seem to feel.

So, what is wrong with WAP? Timing! Everything, if you ask the Nielsen Norman Group[1], who unloaded on WAP when it published a 90-page finding from a field study of WAP users in London. Twenty users were handed a WAP phone and asked to use them for a week. They wrote their impressions in a diary. If you think this calls to mind a certain survivalist series, think again. Health and $1 million were not at stake, but it did seem to put a finger on the weak pulse of mobile Internet vis-à-vis WAP.

When users were asked whether they were likely to use a WAP phone within one year, 70 percent said nay. Nielsen stressed that this finding comes after respondents had used WAP services for a week, so their conclusions are more valid than answers from focus group participants who are simply asked to speculate about whether they would like WAP. Though the report may have been melodramatic about its findings (we surveyed people who had suffered through the painful experience of using WAP, and they definitely didn't like it), it was hopeful for the future.

Mobile Internet will not work during 2001, but in subsequent years it should be big. Nielsen recommends that companies sit out the current generation of WAP, but continue planning their mobile Internet strategy. In other words, don't waste your money on fielding services that nobody will use. WAP usability thus remains poor.

Simple tasks, such as checking weather and reading headlines from their WAP-enabled phones took anywhere from one to three minutes. When one measures that time next to the mere seconds (these functions are processed from a desktop), the discrepancy is resounding.

Maybe users are spoiled by high-speed access, but it doesn't matter. The findings from the Nielsen report make it clear that using a WAP phone is a textbook example of pushing forward back. To make matters worse, cost-efficiency dilemmas are inherent in the process. When one considers that airtime is charged by the minute, it is cheaper to buy a newspaper to check TV listings than it is to spend the minute calling it up on a phone.

Initially, WAP was thought to cause low consumer acceptance of wireless data (and make no mistake—so far, it's a major bomb so far in North America). The WAP forum and Openwave brought a lot of that problem on themselves by trying to turn this technology into a brand name (consumers could care less what enabling technology is used) and raising unreachable expectations for it. The wireless operators compounded the problem by making unmeetable claims in their advertising, such as Sprint PCS' claim to put the Web in your pocket. Users who tried quickly found out that it was no more than a teensy, hard-to-read snippet of the Internet on an unreadable screen with excruciatingly slow access.

Miniaturized Web content is difficult to deal with at best. U.S. Internet users have been spoiled by *low cost, faster Internet access on large color screens*—thus, making it easy to see why wireless Net access was headed for disaster in the United States. For example, based on the preceding discussion, this is where firms such as Phone.com went wrong—by assuming that WAP would be as huge a success as i-Mode is in Japan, where mobile users rely on their phones more than their desktops for Internet access.

So, where does that leave the status of mobile Internet in this country? Is it paralyzed in a stasis of mismatched technology, user demand, and premature business models? Not really.

So, given the sad state of mobile Internet in the United States, is it any wonder DoCoMo was able to buy 16 percent of AT&T Corp.'s wireless unit for $9.8 billion recently? Analysts at Goldman Sachs[2] are extremely bullish on this play, which they indicate could jumpstart the adoption of mobile Internet use in the United States because of DoCoMo's extensive knowledge of pushing i-Mode as a brand. DoCoMo is still going to unleash its next-generation mobile system, based on *wideband CDMA* (W-CDMA), that can support speeds of 384 Kbps or faster, making mobile multimedia possible.

The analysts also indicate that i-Mode does not pose a competitive threat to the embattled WAP spec. Their feeling is that WAP and i-Mode will eventually work together: they believe that the two technologies will converge, with *Extension*

Mark-up Language (XML) offering the underpinnings for the unification between both standards.

Now that it is known that DoCoMo has powered i-Mode's success in Japan, who will bear the WAP torch in the United States? That mantle would likely fall to Openwave Systems Inc.[3] The product of a merger between Phone.com and Software.com, Openwave carries the mobile Internet customers in Japan that NTT doesn't, which could easily make NTT a threat to the new company, whose recent closing has since seen a stock butchery of 50 percent.

Openwave has recovered somewhat since DoCoMo's investment in AT&T Wireless. I-Mode will definitely take a ding out of WAP in the United States now that AT&T Wireless is licensing it from DoCoMo.

Future partnerships are seen by analysts between DoCoMo and Openwave on the horizon, as both firms have expressed interest in XML conversion to wireless. They also see Openwave as a potential technology vendor to DoCoMo, whose Hutchinson subsidiary already moved Openwave's WAP gateway and browser for i-Mode services. Thus, the firms will learn to work together to patch up busted mobile Internet use in the United States.

Now, before going further in this chapter, a detailed definition of WAP is in order. In other words, what really is WAP? Let's take a brief look.

WHAT IS WAP?

WAP is a specification for presenting and interacting with information on wireless (and other) devices. Technically speaking, communications systems consist of many layers (in many cases, seven or more): physical layers (optical fibers, wireless transmission systems, lasers, antennas, and so on), and software layers (transmission protocols and so on). Good engineering practice requires that these layers be decoupled as much as possible. WAP, as a protocol for presenting and interacting with information, is positioned near or at the top of these layers. Therefore, WAP can be used on top of different communication systems. Commercial services can be implemented in many different ways using the WAP protocol. For example, present implementations of commercial services using WAP in Japan and in Europe differ substantially—the user experience is different, commercial models are different, handsets have very different characteristics, and so on. Therefore, it is not useful to identify WAP with one particular implementation. Mostly, the user experience is related to factors that have nothing to do with the WAP specification themselves, but are specific to a particular implementation. In addition, WAP specifications change over time and are coordinated within the *WAP-Forum*, which also produces an excellent WAP.

So, is i-Mode related to WAP? How are i-Mode sites made?

I-MODE'S RELATION TO WAP

I-Mode service is not based on WAP. I-Mode uses a simplified version of HTML, *compact HTML* (cHTML), instead of WAP's *Wireless Markup Language* (WML).

Does it make sense to compare i-Mode and WAP? Let's take a look.

COMPARING I-MODE AND WAP

It's important to keep in mind, that WAP is a protocol, whereas i-Mode is a complete wireless Internet service at present covering almost all of Japan with 28 million subscribers (as of this writing). Therefore, comparing WAP with i-Mode is somewhat like comparing a Rolls-Royce-Jet-Engines with United Airlines or Air France (see sidebar, "Which Is Better: WAP or I-Mode?").

It's not relevant to compare WAP directly with i-Mode. Today, it is relevant to compare WAP with cHTML, or to compare a particular WAP implementation (Japan's EZnet or T-Mobil's WAP service in Germany) with i-Mode. Regarding the future, it is important to look at the role of XML and to follow how the WAP standard evolves.

WHICH IS BETTER: WAP OR I-MODE?

WAP is nothing but a flop. *Wireless application protocol* (WAP), which enables cell phone access to Web content, is over-hyped and overpriced.

The problem is in the Internet markup language. Web sites have historically been grounded in the meta-data system language known as HTML, which isn't compatible with a WAP phone. WAP phones are best served by the WML.

Unless a Web site is written in WML, a WAP phone can't access it. Only 35,000 WAP-accessible sites exist in the world, according to wireless resource Pinpoint.com. However, practically every major telecommunications company worldwide is using WAP as the de facto standard to transmit data on cell phones.

Most users would be inclined to use the Web capability on their phone if U.S. phone companies would adopt a service similar to that of Japanese telecommunications giant NTT DoCoMo's i-Mode, which can read practically any Web page (with varying degrees of legibility) and charges users for the amount of information downloaded rather than air time.

I-Mode is served by cHTML, which technically can enable users to access desktop HTML sites, although it looks better if it's been written in cHTML. Because WAP defines a new markup language, content providers have to learn how to make content with it. This reasoning is one side of what is becoming one of the most contentious wireless Web debates.

Those who agree with the preceding arguments indicate that Japan's i-Mode offers more affordable access rates, more robust content, and higher connection speed. However, through the WAP Forum, WAP boasts the worldwide support of over 600 major phone carriers and manufacturers that have been working together to ensure that their services are compatible with each other. The forum was founded in 1996 by Unwired Planet (now Phone.com).

NTT DoCoMo's i-Mode, meanwhile, is a proprietary service only offered in Japan and can't be made readily available on any other service carrier's network. Many in the industry support it and a lot of phones ship with WAP.

There are no statistics on the number of WAP phones in operation or the number of WAP subscribers worldwide because industry officials and analysts haven't counted. However, market analyst Datamonitor estimates that 27.6 million WAP subscribers are in Europe.

Japan has over 28 million i-Mode subscribers and a little over 4 million WAP users, according to NTT DoCoMo. Officials from NTT DoCoMo and the WAP Forum adamantly deny that the two technologies compete with each other. In fact, NTT DoCoMo is a member of the WAP Forum. However, the two groups still took jabs at each other.

WAP is put together by consensus. It's a democratic process by 600-plus companies around the world.

I-Mode is a specification. If i-Mode was a superior specification or technology, then other companies would have adopted it by now. However, 600-and-some companies have gotten behind the WAP standard rather than the i-Mode standard. That's got to tell you something.

WAP and i-Mode users may agree on one thing: retrieving Web content on the i-Mode is much easier. Before accessing a site, WAP users must agree to pay extra charges and even type in URLs to browse through sites other than the service provider's portal. I-Mode phones have a one-button browsing method, eliminating the need to type in Web addresses.

I-Mode is simpler, but more people are using WAP because it covers so many countries. Also, the transfer speed for i-Mode is fast and WAP is slow at this moment.

The i-Mode offers a lot of content for a fairly inexpensive price. You can send messages relatively inexpensively: for 100 characters, about a cent and a half.

Analysts may love i-Mode, but they admit that NTT DoCoMo is going to have a tough time extending its service past Japanese borders because of WAP's worldwide dominance. Nevertheless, because i-Mode is a proprietary, closed standard, American companies, European companies, and some Asian companies want an open standard where they can have a choice of vendors. WAP is designed to work on any network platform. The technology of i-Mode cannot be deployed on other networks.

Still, to the delight of mobile phone users, NTT DoCoMo has indicated it may offer i-Mode services abroad. Recently, NTT DoCoMo has taken a 15-percent stake in Dutch KPN Mobile and claimed 20 percent of Hutchison 3G (the alliance between KPN DoCoMo and Hutchison Whampoa) to bid for 3G licenses in Europe.

Rumors are circulating that NTT DoCoMo is poised to purchase a 10- to 15-percent stake in the joint venture between BellSouth and SBC Communications. The company is already on its way to completing the purchase of an *Internet service provider* (ISP), Verio

continues

continued

Communications. That deal has moved ahead after concerns that the United States could be left vulnerable to espionage if Japan's government-owned NTT gained access to U.S. wiretapping activities.

Despite the WAP Forum's belief that its protocol will dominate the future of the mobile wireless Web, i-Mode lovers aren't so sure. Companies like Microsoft are bracing themselves to support both WAP and i-Mode.

Right now, this is a young market and many of the standards and market positions are still evolving. Both formats have qualities that the other could benefit from. That's why Microsoft supports both HTML and WAP with Microsoft Mobile Explorer.[4]

NOTE

I-Mode is a brand and a service mark owned by NTT-DoCoMo. As such, i-Mode branded services could employ one or more different application protocols or page description languages in the future.

THE DIFFERENCE BETWEEN I-MODE AND TODAY'S WAP IMPLEMENTATIONS

WAP-based wireless Internet services today are used in Europe, Japan, Korea, and other areas in the world. WAP implementations use a page description language called WML, while i-Mode uses cHTML. Different companies implement wireless Internet services using WAP as a protocol in very different ways. For example, WAP-based services in Japan, which are in competition with i-Mode, provide a very different user experience than WAP-based services in Europe, demonstrating the flexibility of the WAP approach. One important difference from the user and site developer perspective of wireless services is that Web sites for i-Mode are very similar to ordinary TML-based Internet Web sites. I-Mode uses cHTML as a page description language for i-Mode Web sites. Therefore, a very large number of private i-Mode sites are being created. I-Mode sites can also be inspected with ordinary Internet Web browsers (although the result differs somewhat from the display of the same pages on i-Mode handsets). Web sites for WAP-based services on the other hand need to be written in a new and WAP-specific page description language (WML). However, the real business differences between today's WAP implementations and i-Mode are more in the way these services are marketed (advertised business models, charging models, the handsets, battery life, handset display quality, and so on).

Both WAP and I-Mode can package Internet content for mobile devices, but it is not yet clear if either of these opposed standards will come to dominate mobile commerce. Let's see why.

WILL I-MODE CRASH WAP'S PARTY?

NTT *Mobile Communications Network*, known as DoCoMo, is the largest cellular operator in Japan, with more than 30 million subscribers. However, recently, the company has also become Japan's largest ISP by signing up more than 28 million subscribers to its I-Mode service, which offers mobile phone users access to Internet-based information in a similar manner to WAP, which has begun to appear in Europe.

Both WAP and I-Mode deliver cut-down Web pages to mobile devices. Both technologies are in their infancy. The leading applications in Japan are, reportedly, Pokémon screensavers and horoscopes, but the potential of an always-on interactive mobile information service is obvious.

Although both WAP and i-Mode are designed to provide the same service, they are incompatible with each other, and each has its advantages and disadvantages. WAP was developed to be device- and network-independent, offering flexibility and modularity. This has made it more complex, especially because many companies have a significant interest in the technology and have different demands to satisfy.

WAP take-up so far has been low. Indeed, Nokia refused to discuss the number of users it has acquired. They are in the early stages. Expansion will come later.

I-Mode is a simpler solution and is based on HTML. This makes it easier to convert or clip content from Web sites for viewing on smaller screens. Although WAP is designed for *third-generation* (3G) wireless networks, i-Mode is being used on *second-generation* (2G) networks. Analysts indicate i-Mode was simpler to bring to market because it only had to fulfil the requirements of DoCoMo.

Japan has the most developed wireless environment in the world, and i-Mode is very effective. However, its biggest problem is that it is a proprietary technology. Does anything proprietary on the Internet work? No, DoCoMo will have to license it, and it will have to be open. DoCoMo wants to have all its digital subscriber base i-Mode-enabled by the end of 2002, which is around 40 million subscribers. This would make it the largest ISP in the world.

European mobile telecom firm Orange is rumored to be interested in bringing i-Mode to Europe, but this is speculation fuelled by Orange's reputation as an innovator. In an attempt to make an impact outside Japan, DoCoMo has begun discussions with U.S. mobile operators, with the intention to offer i-Mode to U.S. subscribers, who have so far been left out on the mobile commerce revolution.

It is not clear whether i-Mode or WAP will become a dominating standard. WAP seems to have the industry behind it, but i-Mode is more straightforward and has an established subscriber base. Analysts agree that administrators are unlikely to require special provisioning to keep their mobile workers in touch. Enterprises may want to produce specific content for their field workers from their intranet and update it themselves. They may have to think about whether they want to run the server themselves, but it will not make a big difference which way it goes.

So, where in the world are the wireless Internet users? Let's take a look.

LOCATION OF WIRELESS INTERNET USERS

At present (as of this writing), the world's wireless Internet users are distributed approximately as follows (percentages):

- Eighty-one percent of the world's wireless Internet users are in Japan.
- Twelve and one-half percent of the world's wireless Internet users are in Korea.
- Five percent of the world's wireless Internet users are in Europe.
- One percent of the world's wireless Internet users is in the U.S.A.[5]

Also, at present (as of this writing), the world's wireless Internet users are distributed approximately as follows (subscriber numbers):

- **Japan** Thirty million wireless Internet users (i-Mode and WAP)
- **Korea** Three to four million wireless Internet users (WAP)
- **Europe** One to Two million wireless Internet users (WAP)
- **United States** 0.6 million wireless Internet users (WAP and PALM)

Finally, at present (as of this writing), the world's users of WAP-based wireless Internet systems are distributed approximately as follows:

- **Japan** Five million WAP users
- **Korea** Three to four million WAP users
- **Europe** One to two million WAP users
- **United States** 0.3 million WAP users[6]

Which System Has More Users: WAP or I-Mode?

Today (as of this writing), wireless Internet users use the following systems:

- **I-Mode** Sixty-one percent of the world's wireless Internet users
- **WAP** Thirty-eight percent of the world's wireless Internet users
- **PALM** One percent of the world's wireless Internet users[7]

MAIN DIFFERENCES OF WAP IMPLEMENTATIONS IN EUROPE AND WAP AND I-MODE IN JAPAN

WAP implementations in Europe presently (as of this writing) are circuit-switched (users need to dial-up in order to connect). I-Mode is packet-switched (always connected, as long as the user's handset is reached by the i-Mode radio signal). In Japan, WAP implementations also use packed switching. I-Mode includes images, animated images, and color. WAP implementations in Europe, at the moment, only use text and no images.

Great differences are in the business models, charging systems, and marketing of different WAP implementations and i-Mode. For example, in marketing, WAP in Europe is marketed to business; i-Mode is mainly marketed to ordinary consumers.

I-Mode handsets in Japan have large, full color (256 colors) displays and can display animated full color *GIFs* and 10 lines of text or more, whereas European implementations of WAP today have handsets showing four lines of text in black/white without images.

Content marketing of WAP-based services in Europe presently focus on business applications (banking, stock portfolio, business news, and flight booking). In the meantime, marketing of WAP-based services in Japan and i-Mode in Japan focus on fun and everyday life-style: restaurant guides, games, images, and ringing melodies.

NOTE

This is not a limitation of the WAP protocol itself (as Japanese WAP implementations demonstrate), but rather a limitation of present day implementations in Europe. WAP-implementations in Japan do include full color images and many other features not found in Europe at present.

THE FUTURE OF WAP AND I-MODE

It is very difficult to predict the future: who would have correctly predicted the relative development of WAP-implementations and i-Mode three years ago? Future development depends on user/consumer choices, operators' choices and commercial decisions, technical limitations, and even health issues that keep being raised.

Therefore, unexpected developments are not to be excluded. Who would have predicted i-Mode's success a few years ago? Of course, this does not prevent intelligent guesses about the future, but they may well turn out to be wrong. A large factor impacting change will also be the introduction of broadband wireless (3G, UNTS, and so on) services.[8] These will impact the business models, the types of services offered, user acceptance, cash flow models, and so on.

So, why are WAP-based wireless services in Europe implemented as a circuit-switched (dial-up) system, whereas DoCoMo's i-Mode uses a packet-switched system for i-Mode?

WAP-BASED WIRELESS SERVICES IN EUROPE

DoCoMo already had a fully functional packet-switched network installed before introducing i-Mode. This packet-switched network is an overlay over DoCoMo's circuit-switched cellular voice system. European operators at this time (as of this writing) do not yet have such a packet-switched mobile data network. Therefore, they could not roll out a packet-switched WAP service, but needed to roll out WAP services on a circuit-switched (dial-up) basis. The WAP protocol itself has nothing today with circuit-switching or packet-switching. Actually, WAP-based Internet services in Japan are implemented in part on top of a packet-switched network architecture.

WHO WILL WIN—WAP OR I-MODE?

First of all, this may not be a very good question to ask because many different WAP implementations are in the world, and also both the WAP protocol and the i-Mode brand and services offered under the i-Mode brand evolve over time. Second, nobody knows. Third, they may both win. Actually, i-Mode can also be deployed over WAP, or the standards could merge. Fourth, in the future, it is likely (but not guaranteed) that XML encoding will become dominant on the Internet. Therefore, future standards both for WAP and i-Mode could become XML based. In this (likely) case, it is difficult to assign winners and losers. Finally, mobile communications are a revolution, and it's difficult to predict developments.

So, how are users presently (as of this writing) charged by the operators? Let's see.

HOW USERS ARE CHARGED BY THE OPERATORS

In Europe, WAP users are charged for the connection time. For example, if one user looks at a newspaper headline or a football result for 10 minutes, he/she is charged for 10 minutes of connection time. As far as WAP (in Japan) is concerned, different charging models apply to different service offerings.

I-Mode (in Japan) on the other hand is a different story. I-Mode users are charged per packet of downloaded information. So, if an i-Mode user looks at a news item or at a football result for two seconds or three hours on his/her mobile handset, the charge is the same whether it's two seconds or three hours, as long as he/she does not download additional information. In addition to information transfer charges, there is a basic charge and also subscription charges for premium sites, and in some cases, transaction, download, or other charges.

Operating Portals

Finally, in the case of WAP as implemented in Europe, in principle, anybody with an Internet connection can operate a WAP portal—there is the possibility of multiple WAP portals. In Japan's i-Mode, NTT-DoCoMo operates the *official menu* and *i-Mode center(s)*. Anybody can operate an i-Mode site, but he/she needs to enter into a partnership with DoCoMo in order for the site to appear on the *official* i-Mode menu. Only NTT-DoCoMo operates i-Mode center(s).

CONCLUSION

This chapter discussed i-Mode versus WAP—which is better? In conclusion, with wireless Internet usage in the United States falling short of analyst projections, many industry officials are eyeing the wildly successful i-Mode wireless Internet service rolled out by NTT DoCoMo, Inc. in Japan.

I-Mode now has over 28 million active subscribers, each paying an average of $30 per month. I-Mode's success has been scrutinized by Americans partly because Tokyo-based NTT DoCoMo recently bought 15 percent of Redmond, Washington-based AT&T Wireless Services Inc. (see sidebar, "NTT DoCoMo and AT&T Wireless"). The two companies have formed a subsidiary in the United states to focus on streaming media content that can be delivered wirelessly to handsets sometime in 2003, over far faster connections than now are possible.

Although streaming audio and video might seem important only to technophiles who play games, AT&T and NTT DoCoMo officials claimed that rich sound, color graphics, and even streaming video will eventually matter to workers in large U.S. businesses. Salespeople on the road, for example, could use streaming media to offer new sales pitches during important sales calls. Workers could use the service to gain access to graphically-rich corporate intranets.

This will add value to the corporate environment, but the service can't be priced too high for IT managers to accept. Working with AT&T Wireless and NTT DoCoMo to roll out U.S. wireless services over faster bandwidth connections, will be Compaq Computer Corp. and San Mateo, California-based Siebel Systems Inc.

AT&T hasn't indicated how much of i-Mode's look and its business and technology underpinnings will transfer to the United States. However, i-Mode's value is important to both business users and consumers.

E-mail is important to both. When you get e-mail from your boss, you answer it right away. When you get e-mail from your wife, you also answer it right away.

Analysts pointed to i-Mode features and social and environmental factors that fueled its success. In Japan, for example, wired Internet connections are harder to find, slowing the growth of that nation's Web-based home-PC market while making wireless communications a more viable alternative. In the United States, where wired Internet connections are more readily available, there has been less call for wireless Internet access.

NTT DOCOMO AND AT&T WIRELESS

NTT DoCoMo of Japan, regarded as the word's most successful provider of mobile data services, has entered the final stages of talks to acquire a stake in AT&T Wireless. The company wants to buy a 15 percent stake in AT&T Wireless valued at about $10.2 billion. AT&T Corp. plans to spin off AT&T Wireless as part of a restructuring plan announced in the fall of 2000.

NTT DoCoMo has built up a base of over 28 million customers who are ardent users of the company's i-Mode mobile packet data service and wants to launch a similar service in the United States. That follows multibillion-dollar investments in 2000 in mobile carriers in the Netherlands and the United Kingdom.

Besides nationwide coverage for its existing voice and narrow-band data services, AT&T Wireless has another important asset that has attracted NTT DoCoMo: the company has so much spectrum in the United States, that the company didn't need to bid for additional frequencies to support the rollout of wide-band data services.

In order to capitalize on that spectrum, AT&T needs to abandon its current wireless technology, *Time Division Multiple Access* (TDMA), and switch to *Code Division Multiple Access* (CDMA), which better supports high wireless data rates. That would dovetail with NTT DoCoMo's plans, as the company wants to deploy wide-band CDMA services to support high-speed data. Sprint PCS in Kansas City, Missouri, is a nationwide carrier using CDMA technology. It has indicated it can support data speeds up to 384 Kbps without acquiring any additional spectrum.

Nevertheless, the biggest factor in i-Mode's success has been the extensive network coverage. Although the United States has many gaps in wireless service, such gaps are rare in Japan and Europe.

I-Mode has been derided by some Americans as favoring entertainment delivered over flashy color screens, including daily cartoon figures for which users pay about $1 per month. Some critics have indicated Americans might not want a service apparently aimed at a younger crowd.

However, all ages and cultures will be interested in the service. Such cartoons are just one of many features offered by i-Mode. Contrary to some preconceptions, the majority of i-Mode's users are older: just 7 percent are teenagers, 43 percent are in their 20s, 21 percent are in their 30s, and 29 percent of users are over 40.

In response to views that Japanese users have different tastes than Americans, the Japanese are not a separate species. The cultures are very divergent. However, some analysts are not sure that i-Mode can be brought over here.

What is likely to carry over from Japan is the heavy use of packet-based billing. Under that billing model, users are charged by the number of packets they receive over their phones, a process designed to help lower costs.

NOTE

When NTT DoCoMo runs a service on its phones, it charges a 9-percent commission for doing so.

I-Mode has also helped its own cause by sharing revenues with the developers of the services it provides, such as mapping, train information, weather, and stock prices. If the telcos don't make it profitable for wireless developers to do their work, the wireless data service in the United States will not succeed.

END NOTES

[1]Nielsen Norman Group, 48921 Warm Springs Boulevard, Fremont, California 94539-7767, 2001.

[2]Goldman Sachs & Co., 85 Broad Street, New York, NY 10004, United States of America, 2001.

[3]Openwave Systems Inc., 1400 Seaport Boulevard, Redwood City, CA 94063 USA, 2001.

[4]Wired Digital, 660 Third Street, Fourth Floor, San Francisco, California, 94107, USA, 2001.

[5]Eurotechnology Japan K. K., Parkwest Building 11th Floor, 6-12-1 Nishi-Shinjuku, Shinjuku-ku, Tokyo 160-0023, Japan, 2001.

[6]Ibid.

[7]Ibid.

[8]John R. Vacca, *Wireless Broadband Networks Handbook: 3G, LMDS, and Wireless Internet*, McGraw-Hill, New York, 2001.

CHAPTER NINE

I-MODE EMULATORS: TESTING

A number of i-Mode emulators exist. Some are Web-based; others are standalone. NTT-DoCoMo's Paris subsidiary also has a *DoCoMo-Europe's i-Mode demonstrator* with very limited functionality on its Web site (**http://www.docomo.fr/eng/welcome/default0.htm**).

WHAT THE TYPICAL I-MODE SCREEN LOOKS LIKE

Figure 9-1 shows separate screens of Eurotechnology-Japan Corp.'s **http://www.eurotechnology.com/i/i-Mode** site on a N502it (left) handset, and a NM502i (right) handset.[1] You should note that DoCoMo has 80 or more different handsets (counting color variations) and many more for competing mobile Internet systems in Japan.

In other words, i-Mode displays are somewhat larger than regular cell phones. Some models are monochrome, whereas others display gray scale or 256 colors. Most models can show small animations (animated GIFs). The size ranges (as shown in Figure 9-2) from the smallest screen with 96 × 108 pixels (D501i) to the largest one with 120 × 130 pixels (N502i).[2] This corresponds to anywhere from 6 to 10 lines of text, at 16 to 20 characters per line.

imode screen (256 colors)

imode screen (black & white)

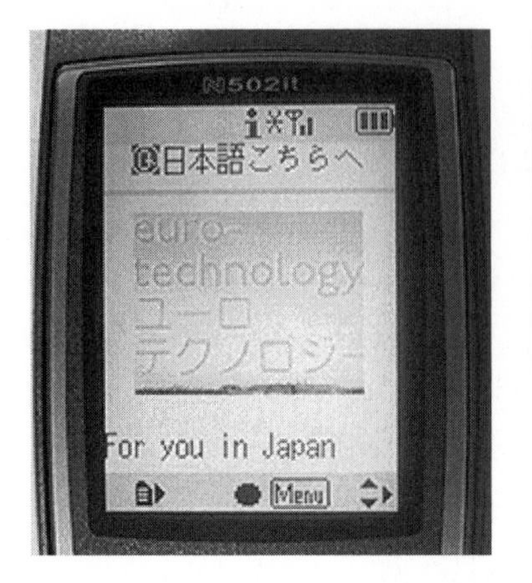

FIGURE 9-1 The N502it (left) and a NM502i (right) handsets.

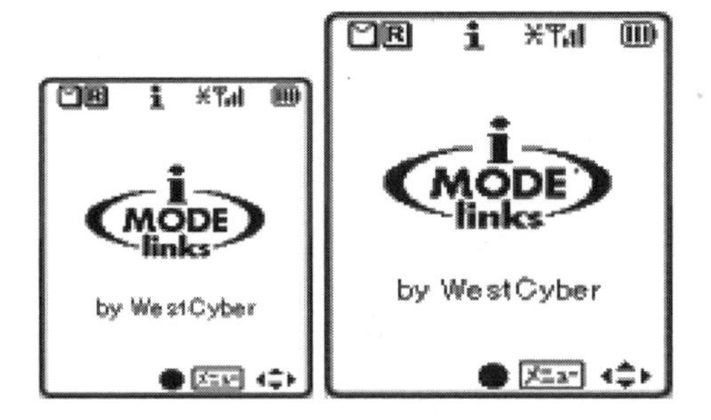

FIGURE 9-2 I-Mode handset display sizes.

USING AN I-MODE EMULATOR TO DEVELOP I-MODE CONTENT

Of course, you can use an i-Mode emulator to develop your content. However, this is not a very good way to develop i-Mode content. One reason is that in order to develop i-Mode content for Japanese consumers, you need to be in touch with Japanese users. So, you need to have a feel for their needs as well as for the key touches needed to reply, and so on. Also, i-Mode has many different handsets. Some have only black-and-white screens, others have color, others enable only still GIF images, whereas many display animated GIFs. In addition, the line breaks appear differently on different handsets. In order to develop successful content, you really need to test your content on real i-Mode handsets. Because these are so easily available in Japan, it's difficult to see the use of emulators in Japan.

LOOKING AT I-MODE WEB PAGES WITH ORDINARY NETSCAPE OR INTERNET-EXPLORER BROWSER

Of course, because *compact HTML* (cHTML) is an extended subset of HTML, you can use your Netscape or *Internet Explorer* (IE) browser to look at i-Mode pages (see Table 9-1).[3] However, at the moment, 99.999 percent of i-Mode users are Japanese and therefore almost all i-Mode content is in the Japanese language. Therefore, you will need a Japanese-enabled browser.

You will also not be able to see i-Mode-only tags, such as the links that dial a telephone connection directly from the i-Mode handset in Japan. You can find a list of i-Mode-compatible HTML 2.0 tags at **http://www.nttdocomo.com/i/tag/lineup.html**.

You will also not be able to see the many special DoCoMo i-Mode symbols. See the list of DoCoMo's special symbols at **http://www.nttdocomo.com/i/tag/emoji/index.html**. These will usually be replaced by a question mark. So, looking at an i-Mode page with an ordinary PC-based browser will give you an idea, but will not exactly reproduce what i-Mode users see on their handsets:

Return to iraode-FAO main pa oe

CONCLUSION

This chapter discussed the testing of i-Mode emulators. In conclusion, there appears to be only one i-Mode emulator, and it is

TABLE 9-1 I-Mode Browsers and Emulators

PRODUCT	PRODUCT DESCRIPTION
Access (Japan) Access (United States)	Based on a subset of HTML 4.0 standard, Compact NetFront extends the existing Internet infrastructure to support the next generation of mobile information appliances with memory, display, and bandwidth constraints. With NetFront embedded in cellular phones, users can access various services, such as weather forecasts, transportation schedules, data searches, and news updates, anytime and anywhere.
C.media	C.media i-Mode simulator.
Ezos	CHTML/WAP/xHTML browser.
I-browser	I-browser i-Mode browser for Windows.
I-Mode	I-Mode browser.
I-Emulator	Java i-Mode emulator.
Seiichi Nishimura	I-Tool i-Mode emulator emulates D501i/F501i, P501i, and N501i (in Japanese).
NTT DoCoMo	I-Mode World i-Mode demo.
OpenWave (formerly phone.com + software.com)	I-Mode support announced January 2001.
Pixo	Pixo Internet microbrowser. Supports standard HTML and cHTML.
X9	I-Mimic i-Mode emulator.
Zentec	I-Jade; A platform that emulates an i-Mode cell phone with built-in KVM on an ordinary PC, enabling the user to debug applications during development for target cell phones and retain the Java development capital.

in Japanese. How NTT DoCoMo is planning to take over the world with an i-Mode emulator that only supports Japanese is beyond everyone. Still, it is possible to use with a Latin alphabet.

The emulator runs in the Windows9x environment and requires the Visual Basic Runtime SP3 extras. To get the i-Mode emulator up and running, you first need to download the emulator itself. Although programmers do not usually host software not written by them on their sites for copyright reasons, they've made an exception this time. First of all, it was impossible for a programmer to read any copyright notes in the application because of the Japanese language.

In other words, you need to download the emulator itself, and that can be found at **http://allnetdevices.com/faq/extras/i-Mode_sim.zip**. Then, you need the Visual Basic Runtime extras that can be found at **http://allnetdevices.com/faq/extras/vbrt63_3.exe**.

Now, unpack the i-Mode_sim.zip archive to a directory of your choice. Then, unpack the vbrt63_3.exe file (self-extract) to the same directory. Next, run Setup.exe found in the same directory. You'll now start to notice the strange characters that your Windows cannot render. Just click straight through the installer. Finally, launch the emulator by running I-Tool.exe.

END NOTES

[1]Eurotechnology Japan K. K., Parkwest Building 11th Floor, 6-12-1 Nishi-Shinjuku, Shinjuku-ku, Tokyo 160-0023, Japan, 2001.

[2]WestCyber Corporation, 1350 Broadway, Rm. 705. New York, NY 10123, 2001.

[3]palowireless.com, Bogan Street, Summer Hill NSW 2130, Australia, 2001.

CHAPTER TEN

SECURITY ON I-MODE

As you know, NTT DoCoMo's i-Mode is widely used in Japan and is branching out to Europe and the United States. It is also providing serious competition for WAP.

Also, as previously explained, i-Mode is a proprietary service that enables users to connect directly to the Internet using *compact HTML* (cHTML). It is a packet-switched service and is *always on*. Users are charged only for downloading data. Most other wireless networks are circuit-switched, and users are charged for connection time.

Information about security in i-Mode is very sketchy, and security is important. However, why is security an issue on i-Mode?

WHY SECURITY IS AN ISSUE ON I-MODE

Mobile commerce (m-commerce) is conducted on i-Mode, including mobile banking and security trading; therefore, security is a serious issue.

The following are the kinds of security issues found on i-Mode:

- Security of the radio link between the i-Mode handset and the cellular base station (this link uses proprietary protocols and encoding controlled by NTT DoCoMo).

- Security of the transparent public Internet connection between i-Mode sites and the handset in the cHTML layer
- Security of private networks on i-Mode
- Security of private network links between the i-Mode center and special service providers, such as banks
- Password security

The security issues on i-Mode are divided into different sectors, as shown in the following sidebar. Each of these different security issues needs to be addressed separately.

One message from a recent mailing list indicated that because NTT would not publish any information about i-Mode security, it should not be considered encrypted or secure. Without more details, what choice does anyone have?

NTT is aggressively pursuing partnerships in Europe and the United States, most recently with AT&T Wireless. There is no guarantee that i-Mode will meet with the same success in Europe and the United States that it enjoys in Japan, but it would be sad to see it become the choice of the people if it does not have adequate security. Users will end up patching i-Mode the way they have patched thousands of networks before.

Another school of thought is that in the future, the two (WAP and i-Mode) could join forces to work out a new standard, where they would be compatible. A step in this direction could be the fact that NTT DoCoMo has become a very senior WAP forum member, and the next version of WAP could be a combination of the two (see the sidebar, "Security Issues in WAP"). So, a new standard might emerge where the two will be compatible, and this means good things for technology as well as for mobile users.

SECURITY ISSUES IN WAP

Early in 2000, the type of content available to wireless users was limited mainly to weather, news, sports, stock quotes, and so on. The nature of this informational data did not require robust security measures, and the attainable level of security was not high enough for most banks and brokerage houses to offer their online services to wireless customers.

Although the WTLS specification within WAP provided strong security between a WAP client and the gateway, there was no way for WAP to interface directly with the Internet, and all WAP traffic had to be translated to HTTP in the gateway in order to reach content servers on the Web. This became known as the *gap* in WAP. Anything that was encrypted with wireless protocols had to be decrypted in the gateway, and then reencrypted with Internet protocols, and vice versa.

To bridge this gap, the WAP Forum developed additional specifications and now can offer the end-to-end security that will enable wireless e-business and banking. The following are two functions that will be necessary to conduct banking and other commerce activities on the wireless network: the capability to exchange electronic documents, and the capability to make secure payment transactions.

In the first case, strong user authentication, integrity and confidentiality of the data exchanged, and end-to-end security between the service provider and the end-user are key in the deployment of the service. In the second case, a nonrepudiation mechanism for the payment transaction, acceptance of the security scheme by the banking organizations (Visa, MasterCard, and so on), and again, end-to-end security between the payment authority and the end-user, are essential for the deployment of the service. The *Wireless Transport Layer Security* (WTLS) specification provides for three modes of operation:

- Privacy and data integrity only
- Privacy, data integrity, and WAP gateway authentication
- Privacy, data integrity, WAP gateway authentication, and WAP client authentication

Note: WTLS is the security layer of the WAP, providing privacy, data integrity, and authentication for WAP services. WTLS, designed

continues

continued

specifically for the wireless environment, is needed because the client and the server must be authenticated in order for wireless transactions to remain secure and because the connection needs to be encrypted. For example, a user making a transaction with a bank over a wireless device needs to know that the connection is secure and private, and not subject to a security breach during transfer (sometimes referred to as a man-in-the-middle attack). WTLS is needed because mobile networks do not provide complete end-to-end security. WTLS is also based on the widely used TLS v1.0 security layer used in the Internet. Because of the nature of wireless transmissions, modifications were made to the TLS v1.0 in order to accommodate for wireless' low bandwidth, datagram connection-limited processing power and memory capacity, and cryptography-exporting restrictions.

The similarity between WTLS and SSL/TLS is no coincidence, as WTLS was based on *Secure Sockets Layer* (SSL) v3.0 (TLS v1.0). Most SSL implementations that you are familiar with on the Internet require server authentication to ensure end-users that they have not accidentally clicked their way to some rogue Web site. Client authentication is generally accomplished via user ID and password, and not by digital certificate. Many reasons exist why the typical Web user is not likely to have a certificate. Among them is the user's ability or inability to keep a private encryption key private.

Smartcards have long been proposed for holding digital certificates, private keys, and other means of proving one's identity, but, in the United States, they have not been widely deployed. One reason a company might be reluctant to embrace the technology is the trouble and expense of retrofitting the existing corporate infrastructure with smartcard readers. The adoption of smartcards has been anticipated for years, but has been slow in coming, and the trend is likely to continue for corporate PCs.

Wireless phones, on the other hand, and other handheld devices may soon come equipped with a slot for a smartcard, which will make the adoption of the technology much easier and much more acceptable in the wireless world. Two new specifications in WAP v1.2 and higher address these limitations: WMLScript crypto library specification and *WAP Identity Module* (WIM) specification.

The WMLScript crypto library provides the means for cryptographic functions to be initiated on the WAP client from the content provider's Internet Web site. It is an end-to-end security mechanism.

Within WAP v1.2 and higher, the WMLScript crypto library provides the digital signature from the WAP client and will provide other functions (like encryption/decryption of small parts of data or verification of signed contents) in future releases. This first function within WAP v1.2 or higher has a major importance because it solves the nonrepudiation issue for the WAP client side.

WIM specification defines the WIM as a tamper-resistant device and as an independent smartcard application. It is clear that implementing the WIM functionality on a smartcard is both appropriate and desirable. The WIM works in conjunction with WTLS and provides cryptographic operations during the handshake. It is also used for securing long-living WTLS secure sessions.

A tamper-resistant smartcard is an appropriate place to store and protect permanent, typically certified, private keys. The WIM uses these keys for such operations as signing and key exchange. The private keys never leave the WIM.

These three specifications: WTLS, WMLScript crypto library, and WIM act together to provide the desired foundation for conducting e-commerce[1] transactions across wireless networks: privacy[2], data integrity, authentication, nonrepudiation, and end-to-end security. Other ways of eliminating the gap in WAP are also under consideration, including the use of proxy servers.

The WAP Forum, a coalition of vendors that support the standard, indicates that WAP v1.3 and higher has and will eliminate the WAP gap via a client-side WAP proxy server that communicates authentication and authorization details to the wireless network server. WAP v1.3 is scheduled for public release sometime in 2002.

It is encouraging to see that security has been a major initiative of the WAP Forum right from the beginning. Isn't this what you've wanted for after seeing so many security afterthoughts get patched into one network after another? Somebody is finally doing it right, but is it enough?

WAP is not without its critics, and many say that it is only a temporary solution that will fade away in a couple of years. WAP is seen as a *tactical* technology that will be absorbed by 2004 as the poor bandwidth and latency deficiencies of mobile networks are resolved. In addition, Java will be one of the technologies that sidelines WAP because it provides the superior usability needed for functions such as maps, device independence, and crucially, better gaming opportunities. As more sophisticated handsets come to market, WAP will be seen as the lowest common denominator for service delivery.[3]

I-MODE USER AUTHENTICATION WITHOUT COOKIES

For an application that needs to provide some form of user authentication or access control, you need to implement some kind of security mechanism. The i-Mode phone terminals do not support HTTP cookies at this time. This means that the standard application-centric storage (ACS) authentication mechanism (**http://www.arsdigita.com/doc/core-arch-guide/security-and-sessions**) will not work. However, cookies are merely one way to store a small amount of state on a browser and have it returned to the server with subsequent requests. You can also pass state in the URL path or query string. So, it is possible to augment the ACS security system to store its authentication tokens in the URLs, although this requires some modification to the ACS request processor in order to work transparently.

I-Mode phones can, under certain circumstances, transmit their phone number to the Web server, but currently, this information is only sent to Web servers that are registered by NTT and presumably pay a fee for this information. If you are running a for-pay service and want to use NTT to do the billing of customer phone accounts, then this is something you need to do. Otherwise, you get no unique identifying information in the HTTP headers from the phone. The only headers you get from an i-Mode browser are Host and User-Agent as shown:

```
Host: hqm.arsdigita.com
User-Agent: DoCoMo/1.0/D209i/c10
```

Basic HTTP Authentication

I-Mode phones do, however, support HTTP basic authentication (**http://www.w3.org/Protocols/HTTP/1.0/spec.html#AA**). This is a simple challenge-response authentication system where the user is prompted to enter a username and password in the dialog box in the browser terminal, and these credentials are sent to the server in the headers with each HTTP request.

NOTE

On i-Mode phones, the password field in the HTTP authentication dialog box that pops up only supports entry of numeric passwords. Normal alphanumeric ACS user passwords cannot be entered this way.

For applications that do not require high security (perhaps the company phone book directory lookup), HTTP basic authentication is quite sufficient. Remember that if you are really concerned about keeping your password secret, you should not be logging in to any Web sites at all over a standard HTTP connection. This applies equally to users of desktop browsers. Anyone logging into a Web site over a nonencrypted HTTP connection is sending his/her username and password in the clear.

For this application, you need to use HTTP authentication. The ACS does not currently have explicit support for this protocol, so you need to define a utility function, *imode_basic_auth,* which takes a username and password and compares these against the credentials sent by the browser.

The utility function imode_basic_auth first checks the headers to see if an authorization header is present. If not, it issues a WWW-authenticate challenge and aborts further script processing. If the authorization header is present, it compares the username and password in the header with those passed into the function as follows:[4]

```
    ad_proc imode_basic_auth {
      { -uname "" -pwd "" }
    } "Compares HTTP BASIC credentials from connection with uname, pwd. You can
leave uname or pwd as emptying
    string for wildcard match. If they don't match, issues
    a HTTP  WWW-Authenticate challenge and aborts script with ad_script_abort.
" {
        set credentials [ns_set iget [ns_conn headers] "Authorization"]
        if {[empty_string_p $credentials]} {
            ns_set put [ns_conn outputheaders] "WWW-Authenticate" "Basic
realm=\"ACS_iMode\""
            ns_return 401 "text/html; charset=shift_jis" "please login"
```

```
            ad_script_abort
        } else {
            set creds [split [ns_uudecode [lindex [split $credentials " "] 1]]
":"]
            set cname [lindex $creds 0]
            set cpass [lindex $creds 1]

            set ok 1

            if {![empty_string_p $uname]} {
                if {[string compare $uname $cname] != 0} {
                    set ok 0
                }
            }

            if {![empty_string_p $pwd]} {
                if {[string compare $pwd $cpass] != 0} {
                    set ok 0
                }
            }

            if {!$ok} {
                ns_set put [ns_conn outputheaders] "WWW-Authenticate" "Basic
realm=\"ACS_iMode\""
                ns_return 401 "text/html; charset=shift_jis" "incorrect login"
                ad_script_abort
            }
        }

    }
```

To use the aforementioned function, put the following call on every page that requires authentication:

```
imode_basic_auth -uname ""  -pwd [ad_parameter
"GroupPassword" imode]
```

For simplicity, because this is a group application, you need to define a single password for all users of this application and leave the username blank. The group password is set in the *[ns/server/yourserver/acs/imode]* parameters section of the site .ini file:

```
[ns/server/yourserver/acs/imode]
GroupPassword=31415
MaxPageSize=5000
```

For an application that requires customization for each user, a variant of the imode_basic_auth function would need to be modified to look up and verify the user's password in the database for each invocation. The user would have entered a username when the first authentication dialog popped up in his/her browser.

You can deal with the fact that users can only enter numeric passwords in the phone's authentication form in two ways: implement a *lenient* password compare, enabling 1 to match A, B, and C; 2 to match D, E, and F, and so on; or require users to have a separate *mobile* password that is all numeric. The first solution requires that you store the user passwords unencrypted in the database so that you can do the comparison. If neither way is acceptable, you could provide a standard HTML form that enables the user to enter their username and an alphanumeric password in normal Text Input fields. If you wanted to use the normal ACS authentication mechanism (security tokens), then that is what you would need to do.

SUPPORTING ACS USER AUTHENTICATION WITHOUT COOKIES

To support security tokens in URL query args, the request processor would need to be modified to scan the URL path or query args for security tokens, in addition to looking for them in cookies. Putting tokens in the path has the advantage that if relative URLs are used, then they will persist from page to page without the application needing to explicitly re-insert them. However, some mechanism similar to the subcommunities proposal would be needed in the request processor to transparently grab this information.

Finally, the i-Mode phones have a number of predefined icons in ROM, which can be accessed by HTML numeric entity references (**http://www.nttdocomo.co.jp/i/tag/emoji/index.html**). For example,  is an icon of a sun shining (see Figure 10-1).

FIGURE 10-1 Predefined icons.

CONCLUSION

This chapter discussed security on i-Mode. In conclusion, NTT DoCoMo has finally begun embedding digital certificates into its cell phones, with an eye to improving security for its 29.5 million i-Mode wireless Internet users.

The company confirmed recently that root digital certificates supplied by VeriSign Inc. and Baltimore Technologies Inc. (see the sidebar, "Baltimore Technologies Provides Security for I-Mode") are built into the 503i series of cell phones, which went on sale at the end of January 2001 and are DoCoMo's first to support Java.[5]

The root certificates pair with private keys held by the certificate authority and together provide authentication of users and Web sites, and enable the use of 128-bit SSL encryption. The use of these functions and the authentication and encryption means i-Mode sessions will now be much more secure when users are accessing Web sites that offer digital certificates.

NTT DoCoMo has yet to suffer any large-scale security problems with i-Mode. Although with user numbers still strongly rising and the number of Web sites also surging, the company is acting now to avoid problems in the future.

BALTIMORE TECHNOLOGIES PROVIDES SECURITY FOR I-MODE

Baltimore Technologies, a global leader in e-security, recently announced that Baltimore e-security technology is being provided to NTT DoCoMo to enable secure mobile Internet commerce in Japan. Baltimore root digital certificates have been embedded into the latest i-Mode mobile devices, the 503i series. This agreement paves the way for Baltimore to offer digital certificates for i-Mode sites and to provide enhanced security for mobile commerce transactions.

With over 29 million i-Mode users and 30,000 i-Mode Web site content providers, Japan represents a huge growth market for mobile commerce services. As the range of mobile services extends beyond specialist information and entertainment, and now to financial services and the mobile intranet, an accelerating demand exists for wireless e-security. This agreement between Baltimore and NTT DoCoMo ensures that mobile phone users have advanced levels of security to communicate and transact with the same confidence as the wired world.

Digital certificates provide authentication of users and Web sites, and also activate the SSL protocol that provides an encrypted channel for communications. SSL, which is the globally accepted standard for Internet security, has been adopted to provide secure communications for the highly popular Japanese i-Mode system.

Digital certificates and the SSL protocol are key components within a *Public Key Infrastructure* (PKI). Baltimore provides wireless PKI systems and products through its telepathy wireless e-security product range. In addition, telepathy provides the core infrastructure required to extend traditional wired PKI to wireless networks. As it is device- and network-independent, Baltimore telepathy can provide client certificates to i-Mode devices.

Finally, the company currently has 29.5 million users on its i-Mode service, which was launched in 1999. Around 2,531 sites are offered by 806 content providers through the i-Mode main menu, while an additional 49,492 sites are operated by other companies and private individuals.

END NOTES

[1]John R. Vacca, *Electronic Commerce: Online Ordering and Digital Money with CD-ROM*, Charles River Media, 2001.

[2]John R. Vacca, *Net Privacy: A Guide to Developing and Implementing an Ironclad Business Privacy Plan*, McGraw-Hill, New York, 2001.

[3]John Schramm, "Security Issues In WAP and I-Mode," SANS Institute, 5401 Westbard Ave., Suite 1501, Bethesda, MD 20816, 2001.

[4]Henry Minksy, "I-Mode Moile Browser Support In ACS," ArsDigita, Headquarters, 80 Prospect Street, Cambridge, MA, 02139, 2001.

[5]"Baltimore Technologies Provide Security to Latest I-Mode Mobile Devices from NTT DoCoMo," Baltimore Technologies, 77 A Street, Needham Heights, MA, 02494, 2001.

CHAPTER ELEVEN

THE INTERNATIONAL I-MODE SPECTRUM

At this point in time (as of this writing), i-Mode services only operate in Japan. However, Dutch carrier KPN is evaluating whether to support rival *Wireless Application Protocol* (WAP) and i-Mode-based technologies in its planned *third-generation* (3G) network (see Chapter 13, "Third Generation (3G) Broadband Mobile")[1] covering the United Kingdom, Germany, France, and Belgium, following its recent alliance with NTT DoCoMo and Hutchinson Whampoa.

KPN[2] intended to deploy Japan's i-Mode wireless Web technology in Europe, either alongside its current WAP services or as a separate application. I-Mode has proven itself as a technology, and KPN is looking for a way to bring i-Mode applications to Europe. At the moment, only a few *Global System for Mobile* (GSM) communications *compact HTML* (cHTML) phones are available, but KPN does not expect that to become a problem.

CHTML, which is currently only used in Japan, is an Internet mark-up language similar to the widely used HTML language. It enables content providers to develop applications quickly without having to rely on WAP converters, and proponents warn that it could spell the end of the road for WAP-based services, which require Web site operators to write WAP-specific versions of their sites.

However, cHTML-based content needs to pass through a cHTML gateway before users can access it on their mobile phones using an i-Mode browser. The underlying network technology does not matter, however.

Logica,[3] for one, released its own m-WorldGate cHTML gateway recently, and KPN is considering whether to adopt this offering, among others. The sudden interest in i-Mode springs from a certain disillusionment with WAP's ability to live up to the hype that has surrounded it since its launch. A recent survey showed that although 50 percent of Web sites now support WAP to some degree, users can only access select streams of text, rather than the full range of graphics, color, and video that they associate with the Internet. However, unlike WAP, i-Mode is able to deliver color images to mobile devices.

WAP critics also claim that it is easier to convert content to run on mobile phones using cHTML, rather than with WAP's *wireless mark-up language* (WML). As a result, i-Mode subscriber numbers have grown rapidly, hitting the 30 million mark recently. Users currently have access to around 30,000 content pages.

INCOMPLETE STANDARDS

The biggest problem WAP technology faces is the new standards that are issued every six months, making it a moving target. That makes WAP hard to work with. It may only take about four months to develop new technology, but it takes between 12 and 18 months to make a commercial application for the marketplace.

WAP downloads will remain slow and expensive until *General Packet Radio Service* (GPRS) provides faster, *always-on* data transmission. WAP takes 20 to 30 seconds to connect. That gets expensive and it's annoying when you are on the move.

However, WAP defenders argue that WML is based on the *eXtensible Mark-up Language* (XML), which has been tagged as the future foundation of the Internet. They also claim that WAP offers better security to customers than i-Mode, because it uses the safeguards of the GSM SIM card.

Platform provider Tantau,[4] however, attested that neither i-Mode nor WAP provided enough security for undertaking transactions. E-commerce[5] companies could only maintain customer relationships by keeping Internet gateways, regardless of which protocol they are based on, under their own control.

WAP and i-Mode can be relatively secure. However, for conversion to and from secure HTTP content, the information needs to be decrypted at the wireless Internet gateway. If that resides with the carrier, it leaves a security hole for transactions to banks and travel companies.

So, which languages are used for i-Mode content? Let's briefly take a look.

LANGUAGES USED FOR I-MODE CONTENT

At this moment in time (as of this writing), almost all i-Mode users are Japanese and most i-Mode services are in the Japanese language. For the G7/G8 meeting in Okinawa, an English menu section was added with limited content (CNN news, Disney, and so on). The English language content is separate and different than the Japanese language content.

I-MODE'S INTERNATIONALIZATION STRATEGY

NTT DoCoMo is at present entering into a number or partnerships in United States, the Americas, Europe and Asia, and Australia. It is possible that DoCoMo will seek to introduce i-Mode-based services, or services similar from i-Mode, to other countries.

The success of DoCoMo's i-Mode service in Japan meant that the personal handyphone system quickly became ignored. Now, Japan and other Asian countries are starting to realize its strengths again.

Who Needs I-Mode?

The *Personal Handyphone System* (PHS) is a technology that can function as a cordless phone in the home, a mobile phone elsewhere, and can handle voice, fax, and video signals (see sidebar, "The PHS Memorandum of Understanding (MoU) Group"). However, the staggering success of the i-Mode service means it has been consigned by some to the dustbin of history.

Now, a resurgence of demand is amongst Japanese business users demanding better voice quality and decent download speeds, while DoCoMo and IBM have launched new consumer data services to meet the renewed demand. Also, it is seeing a new lease of life in China, Thailand, and Taiwan, while Indian fixed-line carriers are considering the technology as a cheap and effective answer for the local loop.

It is quite clear why it is again proving popular. In Japan, it is because of the demand for better quality voice on mobile handsets and assured 64-Kbps data download speeds. PHS offers voice at 32 Kbps. All of Japan's PHS service providers are reporting a steady increase in subscriber numbers after a dramatic fall in 1999.

The user profile is very different now to what it was when PHS was launched. It was marketed as a cheap, mobile service. Teenagers and low-income earners were targeted. However, when i-Mode came along, with all its gimmickry, this demographic deserted PHS. From a high of 8 million subscribers, PHS MoU Group fell to under 6 million. Now they have 8 million subscribers, and the uptake is coming from big city businessmen who demand the better voice quality and fast download speeds offered. For this demographic, a 9.6-Kbps i-Mode service holds few attractions. Especially, as there is a compromise in voice quality."

With all the publicity of wireless Internet, an increasing expectation is that fast services are available. However, they will not arrive until 3G is well established, and that may still be three or four years away. In the meantime, PHS may well be the best solution for mobile data. IBM Japan, for example, recently launched two notepads with built-in PHS terminals, sold

THE PHS MEMORANDUM OF UNDERSTANDING (MOU) GROUP

The PHS standard is a TDD-TDMA-based low-tier microcellular wireless communications technology operating in the 1880-to-1930-MHz band. It is now used by millions of subscribers worldwide in public PHS networks, PHS-WLL networks, corporate indoor PBX applications, and in the home environment. The cell installation architecture, which uses dynamic channel allocation, requires reduced cell station installation planning and costs for the operators, and can be cost effectively deployed in environments ranging from urban to rural environments.

The standard supports a 32-Kbps bearer capability on each of the 24 TDMA frame slots, enabling 32-Kbps ADPCM high-grade speech quality and a variety of data transmission applications. PHS users experience voice quality comparable to wireline service and commercially available operating data transmission speed of 64 Kbps. Soon, 128 Kbps and faster speeds will be available on PHS. The PHS data transmission technology is based on the PIAFS data protocol whose use in existing public networks already occupies 15 percent of the total network traffic and 45 percent of the total calls, continuing rapid growth. Users enjoy high-speed Web browsing, file transfers, and the convenience of e-mail access anytime and anywhere.

Currently, 20 to 25 manufacturers are involved in handset production, with more than 50 models supporting high voice quality and a minimum of 32-Kbps data transmission. Other data transmission applications now popularly adopted include the location identification service, corporate database access, telemetering, and handset-to-handset direct data communication (transceiver).

through PHS operators, and with a pre-registered PHS phone number. DoCoMo indicates that take-up of the notepad has been strong, and itself launched a service to display moving-picture content in mpeg4 format via PHS handsets. Although audio and video streaming will be the main form of delivery, DoCoMo indicates it intends to gradually extend the range of on-demand content from the Internet. Content providers will set the frame rate from an expected average of four to five frames a second to a maximum rate of 14 to 15 frames a second.

The PHS resurgence is not limited to Japan. Extraordinary growth is also in the PHS subscriber base in China and Thailand. Over three million PHS subscribers are in China, and it is expected that this will double in 2002.

The marketing of PHS is completely different in China and Thailand compared to Japan. Instead of a mobile service, it is a cordless telephone offered by fixed-line carriers. China Telecom, which is allowed to offer the service in cities with a population below two million, calls it *personal access service* and essentially uses it as a wireless in the local loop product. Callers pay fixed-line tariffs and can use the service throughout the city. However, unlike in Japan, these users cannot roam out of their hometown.

The top 30 percent of the population is prepared to pay the higher tariffs of mobile services because of intercity roaming. The next 50 to 60 percent of the population can afford the personal access system, although the remainder still cannot afford telephone services. 3G networks being proposed for increased spectrum efficiency can only be sensibly deployed in high revenue markets. Most of Asia will simply not be able to afford them.

In Thailand, it is the eight-channel feature of PHS that is seen as the major attraction. As in China, the service is offered by a fixed-line carrier, Telecom Asia. However, one subscription can support up to eight numbers, so Thai families are buying several $100 handsets and giving each member a separate phone without the costs of a cellular service.

TRANSFERRING EXPERIENCES FROM I-MODE'S BUSINESS MODELS TO OTHER COUNTRIES

Finally, some specific usage patterns and business models may be specific to Japan's circumstances. For example, in Japan, commuters usually spend a long time on trains to go to work or to school. In Europe and the United States, a much higher proportion of workers take their car to work and cannot use their mobile phone for i-Mode while driving a car.

CONCLUSION

This chapter discussed the international i-Mode spectrum. In conclusion, most of us are not enough of a sociologist to tell you if it's a human trait or just an American one, but in the United States, we always seem to be waiting for the knight in shining armor to save us, and that's also true in the wireless world.

The knight errant sits astride a golden steed atop a mountain. The armored horse raises up on its hind legs, pawing the air as the dark clouds and lightning threaten in the background. Finally, he/she charges down the hill to the rescue and, if you look closely, you'll see he's/she's holding the NTT DoCoMo i-Mode phone.

The impression road warriors come back with after traveling to Japan is that the i-Mode service is the answer to all their international wireless problems. Using an open data platform, i-Mode enables any content supplier to easily participate. For users, i-Mode-enabled phones come preloaded with a menu and simple directions. A user follows the prompts that are on the device without having to worry about an *Internet Service Provider* (ISP) or which version of the *operating service* (OS) is installed.

Cost is relatively low because it uses a packet-based system that is always on, and NTT DoCoMo, such as BellSouth in the United States, charges only for the bits downloaded rather than incurring connection charges. The service uses what one road warrior with i-Mode experience calls a *proper* subset of HTML. It's also been called compact HTML, but curiously, the company's Web site never uses that description. Whatever. Everyone says it is easy to set up and easy to use.

A particular i-Mode feature that content providers really like is its *phoneto* tag that will enable a user to place an international voice call to the company hosting the Web page. If you like this idea, make sure you beef up your international call centers first.

I-Mode is like the phone systems long ago, when all you did was ask the operator, and he/she connected you. It's HTML one-dot-zero, with no JavaScript and no cookies.

The percentage of users in Japan accessing the Internet via desktops is one of the lowest among developed nations. To many of NTT DoCoMo's 30 million i-Mode subscribers, a little more than one third of NTT DoCoMo's 89 million-plus customers, i-Mode is the Internet.

Because NTT DoCoMo (the mobile arm of NTT) has little if any competition as a mobile service provider, whatever technology the company develops for the i-Mode service becomes the standard. This makes it much easier for content providers and accounts for the high adoption rates of the approximately 200 new sites coming online weekly.

What's good for business to consumer sites is also good for *business-to-business* (B2B). For example, many companies looking to deploy a wireless strategy to a nontechnical field service force unfamiliar with PC operation will admire its standard platform, its pervasive deployment, and its ease of use, as easy as filling out a form manually, but with the added capability of leveraging networks on the back end automatically.

If you go to the NTT DoCoMo Web site at **www.nttdocomo.com**, you'll see the company's brief instructions: the only requirement for making a Web site viewable by i-Mode terminals is that it can be created using i-Mode compatible HTML.

However, in the spirit of any good monopoly, do you detect a note of arrogance in the concluding comments? Please note that NTT DoCoMo does not offer Web-site development services.

Other than packet-based circuit-based service, the real differences between i-Mode and, let's say WAP, are not in technology, but in execution. In concept, they are quite similar. Business people may admire the ability of NTT DoCoMo to *monetize* content, but that has more to do with the Japanese market, the lower percentage of desktop Internet users, and the headlock NTT DoCoMo has on its own market.

So, to bring everyone back to reality, let's summarize. What you have here is a wireless platform that uses a subset of HTML and requires rewriting Web sites in order to be viewed comfortably on a cell phone display that gives users access to only a subset of the Internet. However, it has a good deal of intelligent user interface design behind it and far fewer, if any, competitors caviling about its shortcomings.

The truth is that NTT DoCoMo is a member of the WAP Forum, although no major support appears outside of Japan for the i-Mode platform. Without the support of the major players in the United States, Canada, Latin America, and Europe, i-Mode, for all of its pluses, will never be pervasive.

Lest anyone thinks knights in shining armor don't make mistakes, NTT DoCoMo Inc. has conducted an investigation to find out why its subscribers are having trouble connecting to the i-Mode service. As a result, the company has taken steps to correct the problem. Sound familiar?

END NOTES

[1]John R. Vacca, *Wireless Broadband Networks Handbook: 3G, LMDS, and Wireless Internet*, McGraw-Hill, New York, 2001.

[2]Royal KPN N.V., P.O. Box 30000, 2500 GA The Hague, The Netherlands, 2001.

[3]Logica, Inc., 32 Hartwell Avenue, Lexington, MA, 02421, 2001.

[4]724 Solutions Inc., World Headquarters, 10 York Mills Road, 3rd Floor, Toronto, ON Canada M2P 2G4, 2001.

[5]John R. Vacca, *Electronic Commerce: Online Ordering and Digital Money with CD-ROM*, Charles River Media, 2001.

CHAPTER TWELVE

E-MAIL, SHORT MESSAGING, MESSAGE FREE, AND MESSAGE-REQUEST

A number of different e-mail related services are on i-Mode: e-mail, short messages (similar to *short message service* [SMS]), message-free, and message-request. E-mail on i-Mode is just like ordinary Internet e-mail, and i-Mode users can send and receive e-mail to and from any Internet e-mail address in the world. There are a few limits: i-Mode e-mails have to be shorter than 250 Kanji (double byte characters), or shorter than 500 Roman characters (single byte characters). If you send an e-mail longer than this to an i-Mode user, DoCoMo's i-Mode center will cut off any part of the message beyond this limit and attach a warning notice to the user notifying the user of this deletion. Also, attachments are not sent to the i-Mode subscriber, but are discarded at the i-Mode center. So, the rule is: e-mails to and from i-Mode handsets need to be shorter than 250 double byte characters and shorter than 500 single byte characters and no attachments.

The default e-mail address of i-Mode users is 090xxxxxxxx@docomo.ne.jp, where "090xxxxxxxx" is the mobile telephone number. However, the user can change this default e-mail address anytime to a different one, such as his or her name as long as it is not taken by anyone else.

READING YOUR REGULAR E-MAIL ON I-MODE

You can read your regular POP3 e-mail on i-Mode. The easiest way is to set your regular POP account to forward your e-mail to your i-Mode e-mail address that consists of your mobile phone number followed by @docomo.ne.jp (example: 09074065370@docomo.ne.jp), if you haven't changed it to something like *yourname@docomo.ne.jp*.

If your POP3 mail provider doesn't enable you to automatically forward all your mail to another address, or if you prefer to access your POP3 mail only from time to time on your i-Mode phone, you can alternatively use a POP3 *mail gateway* service like Netvillage's *Remote-mail* that enables you to remotely access your POP3 mails from your i-mode phone.

SIZE OF AN I-MODE E-MAIL MESSAGE

Using an i-Mode phone, you can send and receive e-mail messages with up to 250 (double-byte) Japanese characters or 500 Latin characters in the body of your message, including spaces (total allowed size: 500 bytes). If an e-mail message is bigger than that, all text after the first 250 characters will be cut off without warning and cannot be read.

SENDING AN E-MAIL WITH AN ATTACHED FILE FROM A REGULAR COMPUTER

If you try to send e-mail with an attached file (an image or a text file) to an i-Mode phone, the attachment will be deleted by DoCoMo before the message is delivered to the receiver, with a remark at the top in Japanese that says the attachment has been deleted. So, if you would like to send an image to an i-Mode phone (to use it as a phone display wallpaper), you will have to upload the image first onto a Web server and then e-mail the URL for that image instead of the image itself. When the phone user sees the URL in his/her e-mail and clicks on it, the phone will connect to the URL and display the image or Web site.

NOTE

I-Mode phones only support GIF images and cannot display JPEG or PNG images.

NOTE

It is not possible at this time to access your i-Mode e-mails (your_name@docomo.ne.jp) from a computer or other device.

THE NUMBER OF USERS THAT USE I-MODE E-MAIL

E-mail is the most popular application of i-Mode. Users can switch off e-mail (if they don't want to use e-mail), but most users send many e-mails every day.

USING E-MAIL ON I-MODE

The precise method on how to use e-mail will depend on your i-Mode handset. However, normally, you will need to press the i-Mode button on your i-Mode-enabled handset. A menu will appear, and one of the menu items will be *mail* (in Japanese or English characters as the case may be). You then select mail and follow the instructions.

SHORT MESSAGE SERVICE (SMS) ON I-MODE

The SMS service on DoCoMo's phone system is somewhat different than the e-mail service on i-Mode. SMS service can be sent directly to other DoCoMo phones, which subscribe to DoCoMo's SMS service, but that do not necessarily need to subscribe to i-Mode (which is still the majority). The limits and characteristics of SMS messages are somewhat different than e-mail on i-Mode (see Table 12-1).[1]

TABLE 12-1 SMS Development Tools

Tools	Tool description
3G Lab	Alligata is the world's first professionally supported open source WAP and SMS gateway for the mass commercial market.
alphapaging.net	Text messaging and pager software for cell phones and pagers.
ANAM	ANAM delivers an enterprise wireless Internet product that is specifically designed for Windows NT and 2000 server platforms. The wireless Windows product includes a combined WAP 1.1 and SMS gateway, and boasts an excellent price/performance metric.
Autopage	Giving you the ability to route messages between the PCs and legacy systems in your office and all GSM phones. You can also send messages to most pagers. SMS to and from Orange, Vodafone, Cellnet, One2One, and all GSM systems.
Converse	The Converse Mobile Internet Division supplies service-enabling platforms for wireless data solutions plus value-added services designed to nurture customer loyalty and increase revenues. Sharing the stage with their flagship product, the *Intelligent Short Message Service Center* (ISMSC), are the InfoPeeler™ wireless portal and the CellCaster™ cell broadcast center.

Tools	Tool description
Deltica	Deltica's SMS gateway service enables you to send SMS messages (supports OTAP to Nokia 7110) from your Web server or e-mail software to mobile telephones anywhere in the world. Per-message pricing—no setup costs.
DeSoft	DeSoft pager and SMS wireless text messaging software. SMS Center is a 32-bit (Windows® 95/98/NT4/2000) client for sending SMS messages to a mobile phone on any of the four major UK Network networks or SMS/paging messages to any worldwide network whose dial-in service center supports the TAP protocol.
Dialogue Communications	Content delivery over SMS and WAP, wireless e-mail (SMS, WAP, and GPRS), and multimedia messaging.
Digital Mobility	Wireless Cube plug and play wireless SMS messaging and WAP gateway solution. The Wireless Cube supports a powerful family of SMS messaging and WAP services, enabling organizations to roll out wireless messaging extensions to their own corporate networks quickly, efficiently, and economically.
Dynamical Systems Research	SMS gate SMSgate servers, SMSmail Psion software.
email2SMS	Simple utility to automatically collect e-mail and forward some details to a mobile phone via SMS.
Empower Interactive Group	**SMS gateway** A high-capacity corporation to carrier grade IT solutions for complete SMS messaging functionality.
EverMore Technology	Solutions for GPS-enabled location-aware devices. Chipsets and modules suitable for a variety of GPS applications, such as car navigation, vehicle locating, fleet management, and time reference for telecom systems.
EViato	SMS-server suite (reliable high performance SMS) gateway API in Perl, Java, C, Distributed Design, Posix compatible, support Linux, inbound SMS.

continues

TABLE 12-1 SMS Development Tools *(continued)*

TOOLS	TOOL DESCRIPTION
GEI Hamon Systems (New!)	Tools include auto document converter, corporate tools, SMS e-mail, system administration, and Microsoft Outlook Sync Manager.
GPA Technologies	SMS gateway is a 32-bit Windows utility that enables you to send and receive text and binary *short messages* over GSM digital cellular telephone networks.
Kannel	Open source SMS gateway.
Kuulalaakeri SMS-gateway	Kuulalaakeri SMS-gateway is powerful Unix-based platform for operators and service providers. SMS-gateway supports CIMD, EMI, SMPP, and Sema SMSC-protocols. Easy API (HTTP) offers fast way to develop one- and two-way SMS-services.
Jataayu	SMS gateway. Short messaging capability in the shortest time.
Lucent Communications Software	Short message service center solutions.
mail2sms	A filter that converts a (large) mail to a tiny text with contents from the mail.
Mark/Space Softworks	Wireless messaging software. SNPP send via TCP and SMS send/receive via a GSM Phone.
mi4e	ThunderSMS™ Send component sends text SMS messages to any SMS compatible phone. It can be used with all popular Windows programming languages, including C++, Visual Basic, or ASP. Sample Visual Basic and ASP program code is included.
Nethix	WAP server and SMS server toolbox for embedded systems.
NetInformer	SMS wireless messaging service to send SMS to any mobile phone around the world. Can also receive a customer SMS and immediately respond on your behalf.

Tools	Tool description
Network365	mZone mobile commerce server for WAP, i-Mode, and SMS.
NotePage	Alphanumeric paging and wireless messaging software. NotePage's software products are designed to work with alphanumeric pagers, numeric pagers, PCS, cellular and digital phones, PCMCIA pager cards, billboards, and other paging devices. Includes NotePager, PageGate, and WebGate.
PCSInnovations	mobileMAGIC is a server-based software platform providing developers with a tool to develop and connect their applications to their customers on a wide variety of wireless networks, infrastructure, and terminal equipment.
Peramon Technology	SMS server and WAP tools.
ProduktivData	PSWinCom component suite is a flexible ActiveX tool for integrating SMS (GSM) and paging support into your applications.
Prosteps	ActiveSMS—Send and receive SMS messages using the ActiveX component or Native Delphi VCL.
SAMSEM	A simple SMS interface programs for GSM devices.
Simplewire	Simplewire is a wireless text-messaging platform capable of messaging across all wireless networks regardless of carrier. The products and services simplify and enhance the way businesses and consumers communicate through all SMS phones, RIM devices, one- and two-way pagers.
SMS Client	A client implementation for the Cellnet GSM SMS center using TAP.
SMSLink	A client/server gateway to the SMS protocol (the short messages sent to mobile phones).

continues

TABLE 12-1 SMS Development Tools *(continued)*

TOOLS	TOOL DESCRIPTION
Txsms	A SMS delivery program that sends mails to GSM cellular phones as SMS text messages
WapMX	WAP Hotmail gateway. POP3 e-mail gateway. SMS server. SMS broker. OTA provisioning, logos, ring tones, vCards, and vCalendar SMS applications. WAP and SMS software development.
XIAM	The XIAM information router enables for the intelligent routing of information between an enterprise and mobile users over SMS. Remote users can easily update a database, reply to e-mails, access an intranet or Internet Web page, or even restart a critical program from any standard mobile device, at any time.
Xsonic	SMS messages from MS Outlook or database applications.

SENDING SMS MESSAGES

If you are a European SMS user and want to send an SMS message to a friend in Japan, how can you do that? The European SMS systems are not directly linked to Japan's mobile message systems. However, i-Mode users who do not explicitly cancel their e-mail connection can receive Internet e-mails directly to their handset. So, as a European SMS user, you can send an e-mail to your friend's DoCoMo handset from any PC with Internet connection. I-Mode users have e-mails, such as xyz@docomo,ne,jp, where xyz is by default set to the full mobile phone number, but can be changed by the i-Mode user.

NOTE

Message-request and message-free are opt-in message services that are either free message-free or paid message-request by the user.

KINDS OF E-MAIL THAT CAN BE SENT ON I-MODE

In the case of the standard e-mail service, which is included in the fee for i-Mode, there are some limits of what types of e-mail you can send. The length is limited to 500 Roman characters or to 250 Japanese double byte characters. If you send an e-mail to an i-Mode destination exceeding this limit, the e-mail will be truncated, and there will be a short warning attached to alert the recipient that the e-mail was truncated. Also, you cannot send attachments to i-Mode destinations. The attachments will be discarded by the i-Mode center, and there will be a short message to the recipient that the attachment was discarded. There are also limits on how many e-mails are stored in the i-Mode center, on your handset, and the time limits for the storage time. However, users can subscribe to premium e-mail services with enhanced services.

NOTE

Premium e-mail services are available. Users can subscribe to enhanced e-mail services.

NOTE

DoCoMo also offers mobile e-mail services beyond i-Mode, including e-mail services for images: Users can send each other photographs, and so on.

CONCLUSION

This chapter discussed e-mail, SMS, message free, and message-request. In conclusion, as previously discussed in earlier chapters, NTT DoCoMo, creator of the i-Mode wireless Internet service in Japan, has announced that it will begin offering its i-Mode service in the United States through AT&T

Wireless in 2002. The i-Mode service enables users to wirelessly send and receive e-mail, browse i-Mode-formatted Web sites, and more. Initially, the service is expected to be available in Seattle, with nationwide expansion coming in 2002.

Although the fees for the service have not been officially set, the company is expected to charge $70 for voice and about $17 for i-Mode transmission fees. NTT DoCoMo is also looking into flat rates, something Japan doesn't currently offer for the service.

All of this shouldn't come as too much of a shock to those of you who follow wireless development in the United States.[2] In 2000, NTT DoCoMo bought up 16 percent of AT&T Wireless, so this was definitely in the plans.

It is very exciting to see what all the hype surrounding i-Mode is about. We've all recently heard about how great it is to do everything wirelessly, though this may be because Japan is not wired like we are in the United States, and most people use i-Mode as their main connectivity point to the Internet.

It's interesting timing for AT&T Wireless. Right now, Palm and others are trying to get their wireless connectivity solution adopted by the masses. Meanwhile, AT&T and NTT DoCoMo will now try to infiltrate the masses with their solution.

Seventy dollars for voice and $17 for i-Mode is a little too pricey for the American market. AT&T is expected to drop the price to a flat rate in order to gain the market share when the service first launches. The United States currently has a very slow wireless network. A new network is coming (3G) that will promise very fast connectivity and browsing, but until that time, i-Mode may have trouble catching on.

END NOTES

[1]palowireless.com, 1 Bogan Street, Summer Hill NSW 2130, Australia, 2001.

[2]John R. Vacca, *Wireless Broadband Networks Handbook: 3G, LMDS, and Wireless Internet*, McGraw-Hill, New York, 2001.

CHAPTER THIRTEEN

THIRD GENERATION (3G) BROADBAND MOBILE

Consumer demands for broadband[1] mobile communications at highway speeds will not likely be met by next-generation proposals. Although solutions exist for incremental improvements in capacity and bandwidth, the wireless industry has not yet offered an economical method for providing the broadband channels demanded by mobile consumers on the highway.

This chapter examines a proposed high-capacity infrastructure with moving base stations for providing 3G broadband mobile communication services that are not limited by a user's speed. Moving base stations provide moving communication cells to mobile i-Mode users traveling along a roadway. This 3G broadband mobile solution provides communication services with i-Mode data rates of 20 Mbps or more at any vehicular speed at costs comparable to wireline. In addition to presenting a general technical description of the proposed infrastructure, the following discussion includes a brief overview of wireless 3G broadband mobile technology and obstacles encountered in designing a high-capacity system providing high-bandwidth channels to i-Mode users traveling at highway speeds. However, before we embark on a discussion of broadband mobile communication on the highways of tomorrow, let's define 3G mobile.

THIRD GENERATION (3G) MOBILE

At present, the download speed for i-Mode data is limited to 9.6 Kbps, which is about six times slower than a TSDN fixed line connection. However, in actual use, the data rates are a lot slower, especially in crowded areas, or when the network is *congested*. For 3G, new network hardware and software will be installed, which will increase the data rates by a large factor (in the final stages up to 200 times faster). 3G in Japan will be introduced in stages starting in early 2002. Japan will be the world's first country to introduce 3G (see sidebar, "3G Videophones").

NOTE

The main reason for the relatively early (compared to Europe and the United States) introduction of 3G broadband in Japan, is that Japan has a high density of cellular phone users, and the bandwidth in 2G is running out.

DATA RATES FOR 3G (BROADBAND MOBILE) IN JAPAN

At present, i-Mode data rates are up to 9.6 Kbps, but are usually a lot slower. For 3G, data rates will be up to 2 Mbps (approximately 200 times faster). However, 2 Mbps will only be reached later in time—initially, the data rates will be lower.

NOTE

Initially (spring 2002), 3G will be introduced in Tokyo and Osaka, and later in other areas of Japan. 3G will not be introduced in all of Japan at once.

3G VIDEOPHONES

NTT DoCoMo finally delivered videophone-equipped handsets to participants testing its 3G cellular phone services. The firm had hoped to introduce the technology (which enables i-Mode users to view each other while speaking) in time for initial testing, but software bugs forced a delay.

The P2101V handset, produced by Matsushita Communication Industrial Co., has been delivered to 700 individuals and 500 firms. The video-equipped handset is made possible because of the 3G service's revved-up communications speed.[2]

3G BROADBAND MOBILE COMMUNICATION FOR I-MODE DATA

Commentaries and predictions regarding wireless broadband mobile communications and wireless Internet services are cultivating visions of unlimited services and applications that will be available to the consumer *anywhere at anytime*. Consumers expect to surf the Web, check e-mail, download files, have real-time videoconference calls, and perform a variety of other tasks through a wireless communication link. The consumer further expects a uniform i-Mode user interface that will provide access to the wireless link whether shopping at the mall, waiting at the airport, walking around town, or driving on the highway.

Current wireless infrastructures, however, as well as next-generation proposals, cannot furnish the necessary bandwidth and capacity to provide these services to i-Mode users traveling at highway speeds. Unfortunately, highway travelers will likely be the most demanding of bandwidth and wireless services. A huge captive audience occupying the world's multilane highways will eagerly devour bandwidth to take advantage of time in the car or to enjoy various entertainment services.

Commuters can turn a normally frustrating commute to work into productive time with the appropriate applications and bandwidth. The leisure traveler can access a limitless library of music and travel information or entrance the children with a downloaded movie or computer game. Clearly, a broadband wireless solution is needed to provide mobile i-Mode users the high-bandwidth mobile service they demand at a low cost. The proposed infrastructure discussed next is intended to be implemented as part of a global wireless communication system providing high-bandwidth communication services with a uniform i-Mode user interface independent of the location or speed of the user.

The Wireless Goal

It is becoming increasingly apparent that future mobile wireless communication systems must provide high-bandwidth, low-cost, and reliable mobile services comparable to wireline. The i-Mode user will expect a consistent interface as well as uniform functionality and performance independent of the user's location or speed.

For wireline, *broadband communication* refers to communication of digital information where the information transfer rate ranges from a minimum of 1.544 Mbps (2.048 Mbps in Europe) to a maximum of 155 Mbps (*synchronous optical network*, [SONET], OC-3). Broadband communication will support multimedia broadband applications for the home and office, such as wideband Internet access and information retrieval, videoconferencing, imaging and graphics, high-definition TV, video on demand, stored voice, and other services.

Although wireline technologies can provide the necessary bandwidth and i-Mode data throughput to meet the consumer demands at the home or office, current wireless infrastructures are insufficient to provide high- i-Mode data-rate services in a mobile environment. Furthermore, the proposed next-generation mobile wireless infrastructure will also be severely bandwidth-limited. 3G proposals, which have not yet been implemented, limit the overall i-Mode data rate to 2 Mbps and will not provide more than 144 Kbps at highway speeds.

The Broadband Solution

The most economical and practical infrastructure for providing wireless broadband channels to a high concentration of i-Mode users likely includes a large number of small cells communicating at extremely high frequencies. Systems operating at millimeter-wave frequencies, particularly at 60 GHz, can utilize large sections of continuous frequency bandwidth while exploiting the signal propagation characteristics observed at these frequencies for frequency reuse.

Small Cells

It is clear that for a given service area and a fixed available frequency bandwidth, a wireless infrastructure utilizing small cells can provide more capacity than a system using larger cells. Furthermore, the small cell system can provide larger-bandwidth channels to the same number of i-Mode users in the given service area than the large cell system. Assuming that the capacity of a cell remains constant as its size is reduced, the number of channels that can be provided within the service area increases as the cell size is reduced. The increase in available traffic channels is proportional to the inverse of the square of the decrease in size of the cells of the system.

If the diameter of a cell is decreased by a factor of N, the number of cells that cover the same service area increases by a factor of N2, and the number of available channels within the given service area increases by a factor of N2. Although small-cell systems require a greater number of base stations to cover a given service area than large-cell systems, the infrastructure cost per channel for small-cell systems is significantly less than for large-cell systems.

Millimeter Waves

It is becoming increasingly apparent that in order to provide wireless communications with bit rates in tens or hundreds of megabits, a large amount of bandwidth in the millimeter-wave spectrum range (30 to 300 GHz) must be utilized. In order to efficiently provide high-i-Mode data-rate communications to a large number of i-Mode users, a wide, continuous frequency bandwidth must be used. Antenna and radio transceiver technologies currently limit the total maximum allocated bandwidth to roughly 10 percent of the carrier frequency. Therefore, high carrier frequencies can provide a wide, continuous bandwidth. Many frequency spectrum regulatory agencies around the world, including the FCC, have allocated several large sections of spectrum in the millimeter-wave frequency band. In addition to the high carrier frequencies and continuous spectrum, these blocks of frequency spectrum are particularly attractive for use in a broadband communication system due to propagation characteristics observed in some bands.

Millimeter-wave characteristics dictate short-range, line-of-sight propagation with minimal refraction and reduced interference, while providing a bandwidth capacity approaching coax or optic fiber. These millimeter-wave characteristics require a cellular network topology to be based on a large number of small cells. As discussed previously, the small cells facilitate frequency reuse, resulting in a large number of traffic channels per service area, and thus, high network traffic capacity.

60 GHz

The 60-GHz frequency spectrum is especially suited for a network topology with extremely small cells due to the resonant absorption of electromagnetic energy by oxygen molecules at that frequency. The attenuation of electromagnetic waves at 60 GHz is approximately 15 dB/km. This particularly high electromagnetic wave attenuation present in the 51.4- to 66-GHz frequency band (labeled absorption band A1) facilitates a high rate of frequency reuse in small cells. The high attenuation minimizes cochannel interference in a small-cell system, enabling a particular frequency to be used more often than would be possible at other frequency bands.

Under FCC Part 15, the 59- to 64-GHz band is available for general use by unlicensed devices based on severe propagation losses protected from interference. The FCC stated that the goal was to foster novel broadband communications, and that 59- to 64-GHz offers the greatest potential to enable for the development of short-range wireless radio systems with communications capabilities approaching those now achievable only with coaxial and optic fiber cable.[3] In Europe, 62 to 63 GHz and 65- to 66-GHz are allocated for licensed operation, specifically for the *Mobile Broadband System* (MBS). In Japan, the 59- to 64-GHz band is regulated for use by the MBS. It appears, therefore, that the 60-GHz band is geographically widely available.

THE HIGH-SPEED HANDOFF PROBLEM

Apparently, an infrastructure utilizing small cells in the 60-GHz frequency range can provide large-bandwidth channels to an almost unlimited capacity of i-Mode users. Frequency reuse, coupled with a large-bandwidth frequency spectrum, can be exploited to provide large bandwidth channels. Systems such as this, however, have been suggested in the past and have not gained widespread acceptance due to the fallacy that high-speed mobility cannot be supported in a small-cell environment. Although few will argue that a small-cell infrastructure can provide large bandwidth channels on the order of 100 Mbps to i-Mode users traveling at pedestrian speeds, it seems impossible to most that mobile users traveling at speeds in excess of 60 mph can be accommodated in a small-cell system. These fallacies are based on the observation that as the size of the cells is reduced, the mobile unit tends to cross cell boundaries more often, requiring a large number of handoffs to the point that the calls will be dropped if mobile units are moving at high vehicular speeds.

A PROPOSED BROADBAND MOBILE I-MODE SYSTEM DESCRIPTION

The proposed infrastructure examined in this part of the chapter provides a solution to the high-speed mobility limitations discussed previously. Furthermore, the proposed system can be integrated with other infrastructures to enable high-bandwidth wireless service at a cost competitive to wireline communication. The small-cell architecture of the proposed system enables the use of extremely lightweight low-power i-Mode units that can be used almost anywhere. The proposed infrastructure is especially suitable for high-speed multilane divided highways in urban high-traffic environments. Advantages of cordless and i-Mode systems are integrated by deploying very small picocells along high-traffic roadways. Although each of the picocells has a radius on the order of 100 feet, the system can easily facilitate high-bandwidth communications to i-Mode units traveling at speeds up to and in excess of 100 mph.

This is accomplished by interposing moving base stations between i-Mode units traveling down the roadway and fixed radio ports uniformly distributed along the median of the roadway. The moving base stations enable communication links to be established between the i-Mode units traveling on the roadway and a fixed communication network through the fixed radio ports. As can be seen, the number of i-Mode unit handoffs in this proposed system is significantly reduced from those of conventional small-cell systems because the moving cells provided by the moving base stations track the mobile units.

The moving base stations complete the communication link to a fixed network, such as a *public switched telephone network* (PSTN), through a radio link to the fixed radio ports. The fixed radio ports are interconnected with a fiber optic ring, or a similar signal-transmitting device, to a gateway telephone office and *mobile broadband switching center* (MBSC) connected to the PSTN. Mobile units stopped on the highway or traveling at significantly slower speeds than the majority of traffic are coupled to the wired communication network through fixed base stations.

The proposed highway system is intended to be part of a complete high-bandwidth wireless solution where the same i-Mode units can be used at home, in the office, and while traveling by foot or in a high-speed vehicle. The requirement to provide high-bandwidth channels to a high density of i-Mode users traveling at speeds on the order of 60 mph arises in predictable areas such as highway and train systems. Although i-Mode users in rural areas may demand high-bandwidth channels while traveling at high speeds, the service to these users can easily be provided with larger cells because system capacity is not threatened.

Small-cell systems with fixed base stations can be used to provide high-bandwidth services to pedestrians or fixed wireless users where the density of i-Mode users may be high, but the highest user speed is well below any handoff limitations. The moving base station infrastructure, therefore, is intended to be implemented in areas with a large number of i-Mode users traveling at high speeds, whereas other types of infrastructure are used to provide services in other areas.

Physical Configuration

As shown in Figure 13-1, the moving base stations are arranged along a conveying device, such as a rail, and move in the same direction and at approximately the same speed as the traffic flow along a highway.[4] The conveyor device is implemented in a narrow, elliptical loop such that a series of fixed radio ports are positioned along the median of the highway and between the two long ends of the loop.

Several loops are arranged along the highway and overlap slightly. The moving base stations are spaced apart by a selected distance equivalent to the diameter of the cell served by the moving base station, which is approximately 200 feet. Each of the moving base stations provides a moving cell that travels along the highway in accordance with the i-Mode users it services. The fixed radio ports distributed along the median of the highway between both rails use directional antennas to

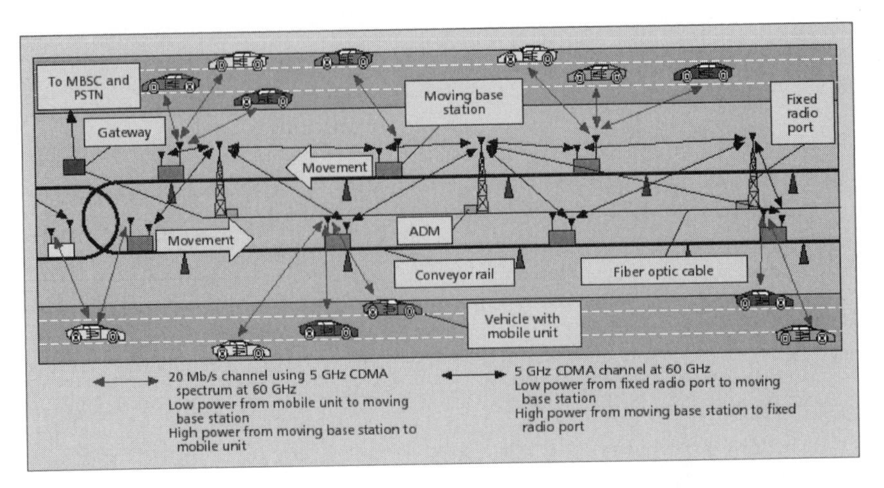

FIGURE 13-1 An i-Mode broadband communication system with moving base stations.

communicate with each of the two groups of moving base stations servicing the two corresponding traffic flows of i-Mode users.

Because the moving base stations track the motion of the mobile units, the relative speeds between the moving base stations and mobile units are less than speeds found in cordless systems. It appears that cordless systems can easily facilitate call handoffs between base stations at pedestrian speeds up to 30 mph. Because all traffic does not travel at the same speed and the moving base stations will travel faster than some of the mobile units and slower than others, the moving base stations can typically handle mobile units traveling up to 30 mph slower or faster than the speed of the mobile base stations. For example, if the moving base stations are traveling at a speed of 60 mph, the moving base stations can provide communications to mobile units traveling at speeds from 30 to 90 mph. Although highway traffic situations rarely involve speed differentials of more than 60 mph, additional features may be implemented to cover those situations.

To avoid interruption in communication, the ends of the loops are sufficiently close or overlapping to provide an overlapping area of coverage for mobile units traveling in the area of loop ends. A handoff procedure is performed to transfer mobile units serving a moving base station nearing the end of a loop to a moving base station on an adjacent loop.

Any one of several techniques may be used to accommodate the different speeds of traffic on the two sides of the roadway. Some of these techniques include using multiple loops or multiple rails and performing mobile unit handoffs to maintain uniform spacing between moving base stations. A particularly efficient method for providing a speed buffer between the two sides of the roadway involves implementing a single additional rail between the two main rails. As shown in Figure 13-2, mobile base stations are directed to and from a center rail as needed. The speed and direction of the mobile base stations traveling on the center rail depend on the speed differential between the two main rails.[5]

A fiber optic cable couples the fixed radio ports to a gateway. *Add/drop multiplexers* (ADMs) facilitate the transfer of signals between the fixed radio ports and the fiber optic cable. The fiber optic cable forms a continuous ring in accordance with SONET or *synchronous digital hierarchy* (SDH) transmission protocols. The gateway provides an interface to an MBSC. The MBSC, which could be implemented as part of the gateway, performs switching functions analogous to a mobile switching

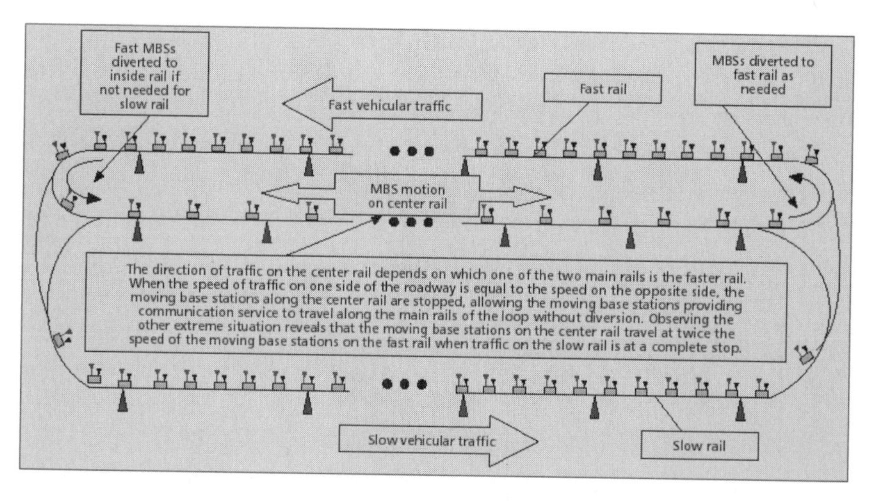

FIGURE 13-2 Moving base station speed management.

center in other types of mobile systems. The MBSC is coupled to a wired communication network such as a PSTN. Preferably, all the fixed radio ports associated with a particular loop are connected to a single gateway. The need for moving base stations to register to a gateway is avoided because the moving base stations only communicate through one gateway and do not need to be handed off to another gateway, as would be necessary in a conventional cellular system.

Communication Links

All wireless communications within the proposed system use direct sequence *code-division multiple access* (CDMA) spread spectrum techniques within a 5-GHz frequency band part of absorption band A1 at 60 GHz. Using *time-division duplex* (TDD) methods, the entire 5-GHz band is used for upstream and downstream communication. In the upstream half of the TDD cycle, i-Mode data is sent from the mobile units to the fixed radio ports by transmitting upstream signals from the mobile unit to the moving base station and transmitting corresponding upstream signals from the moving base station to the fixed radio ports. In the downstream half-cycle, i-Mode data received from the wired communication network is sent from the fixed radio ports to the mobile units by transmitting downstream signals from the fixed radio ports to the moving base station and corresponding downstream signals from the moving base station to the mobile units.

Upstream signals are received from the mobile units at the moving base station and retransmitted to several fixed radio ports using *time-division multiplexing* (TDM) techniques where the communication channel for each mobile unit corresponds to a time slot within the retransmitted upstream signal. Multiple channels between the moving base station and the fixed radio ports are time-division multiplexed as time slots in a i-Mode data stream. The i-Mode data stream is spread with a pseudo-random code over the allocated spectrum. A pilot sequence inserted into the transmitted signal facilitates synchronization using known techniques.

The corresponding upstream signals are received at the several fixed radio ports and forwarded to a gateway through the fiber optic ring. Each fixed radio port receiving the signal from the moving base station determines a quality indicator for each signal based on signal strength and signal-to-noise ratio. The ADMs connecting the fixed radio ports to the fiber optic ring forward the upstream signals and the corresponding quality indicator measurements to the gateway. The gateway processes the signals using the quality indicator measurements to produce a high-quality upstream signal that is provided to the wired communication network.

Downstream signals are received from the wired communication network through the gateway and are directed to the appropriate fixed radio ports based on the location of the moving base station serving the mobile unit intended to receive the downstream i-Mode data. The downstream signals are transmitted through the fiber optic ring and received at several fixed radio ports through the ADMs.

Each signal is transmitted from multiple fixed radio ports in the vicinity of the moving base station serving the mobile unit. The signals are directed to the appropriate moving base station using a unique moving base station address derived from an identification code and a Walsh code. The mobile base station processes and combines the several signals received from the fixed radio ports to produce a high-quality downstream signal. After extracting the downstream i-Mode data from the time slot corresponding to the particular mobile unit channel, the moving base station appends appropriate signaling information to the i-Mode data and retransmits the downstream i-Mode data to the intended mobile unit.

As discussed in the preceding, the moving base stations have one set of directional antennas aligned to service moving traffic and a second set aligned to communicate with the fixed radio ports. The downstream signals transmitted from the fixed radio ports to the moving base stations are at a relatively low power level, whereas the downstream signals transmitted from the movable base stations to the mobile units are at a relatively high power level. Due to the characteristics of direct sequence spread spectrum CDMA communications, the higher-power-level signals overpower the lower-level signals such that the mobile units do not receive communications from the fixed radio ports and only receive those signals transmitted from the movable base stations.

Upstream signals are transmitted at a relatively low power level from the mobile units to the moving base stations compared to the upstream signals transmitted from the base stations to the fixed radio ports. Any direct communication between the mobile units and the fixed radio ports therefore is minimized. Accordingly, the 5-GHz spectrum is efficiently used by the radio interface between the moving base stations and the fixed radio ports and the radio interface between the moving base stations and the mobile units. Each radio interface is not affected by the communications within the other interface.

The Moving Base Station-Fixed Radio Port Interface

The radio interface between each moving base station and the fixed radio ports includes 15 channels, each having a 20-Mbps i-Mode data rate. The channels are TDM in 15 time slot frames, resulting in a total bit rate of 300 Mbps. To achieve a processing gain of 9 dB, the frame rate is multiplied by a factor of eight, yielding a 2400-Mbps i-Mode data rate. Due to the TDD communication link, the 5-GHz band has an effective bandwidth of 2500 MHz. If a modulation rate of 1 b/Hz is used, the 2400 Mbps i-Mode data is transmitted in the 2500-MHz effective bandwidth of the appropriate duplex half-cycle. In addition to the 12 channels for bearing communication traffic, the 15 channels include three channels for signaling, control, base station identification, and error coding.

The spacing of the fixed radio ports, together with the strength of the signal transmitted between moving base stations and the fixed radio ports, determines the number of fixed radio ports with which a moving base station can communicate at any point in time. The spacing and signal strength is preferably such that each fixed port receives signals from three moving base stations. The fixed radio port receives the upstream signal that includes the upstream i-Mode data in addition to the address of the moving base station that received the data. A processor in the fixed radio port computes a signal quality indicator for the received upstream signal where the quality indicator is a figure of merit based on signal strength and signal-to-noise ratio. The quality indicator is transmitted to the gateway in addition to the upstream signal through the ADM.

WALSH CODES

The moving base stations are distinguished and identified using predefined code sequences derived using Walsh functions and the identification code for the moving base station. The combination of a Walsh code and the identification code yields the unique address for each of the moving base stations. An eighth order Walsh function provides eight orthogonal codes. The 0 Walsh sequence is used as the pilot carrier with the other seven sequences available for providing communication for the moving base station. Each of the moving base stations therefore has one of seven assigned codes in addition to its identification code.

The Walsh codes are assigned to the moving base stations so that two moving base stations with the same Walsh code will be physically separated by a sufficient distance to prevent interference in communications between fixed radio ports and moving base stations with the same identity code. For example, the codes may be assigned in sequence such that the codes are repeated in the pattern *12345671234567 . . .* to the sequence of adjacent moving base stations arranged along the loop. The Walsh codes are further multiplied by pseudo-noise codes to improve communication performance.

In order to preserve the order of Walsh codes when moving base stations are diverted to the center rail, the moving base stations are only redirected in blocks equal to the number of Walsh codes. For example, if a surplus of moving base stations results at the faster rail, and Walsh codes one to seven are in use, moving base stations are directed to the center rail from the faster rail in groups of seven such that the *1234567* sequence of Walsh codes is maintained for the moving base stations along the loop.

Synchronization

Pilot signals are transmitted from the moving base station to the fixed radio port and from the fixed radio port to the moving base station. The moving base station is synchronized in phase and time to the fixed radio port by phase-locking to the pilot signal. For system synchronization purposes, the moving base station receives a *Global Positioning Satellite* (GPS) *Universal Time Coordinated* (UTC) timing signal once every second.

Upstream Gateway Functions

The gateway receives the same i-Mode data from several fixed radio ports and stores each version of the data in an internal memory in association with the corresponding moving base station address and the address of the fixed radio port receiving the particular version of the data. Accordingly, multiple copies of the same i-Mode data transmitted by a single moving base station are stored in the memory of the processor in the gateway. The signal quality indicators computed by the processor in each of several fixed radio ports are compared to a predefined signal quality indicator threshold. Versions of the i-Mode data corresponding to a signal quality indication below the threshold value are discarded.

A cyclic redundancy code transmitted with the i-Mode data is used to detect any TDM frame errors. The i-Mode data associated with the upstream signal having the best quality indicator is transferred from the gateway to the wired communication network using the appropriate protocols.

DOWNSTREAM GATEWAY FUNCTIONS

The i-Mode data received from the wired communication network at the gateway and intended for a registered mobile unit is stored in the memory of the processor in a register particularly associated with the moving base station currently serving the mobile unit. This i-Mode data is sent through the optical ring to all fixed radio ports that are identified in the memory of the processor as fixed radio ports with an acceptable quality indicator as determined by the previously received upstream signals. The received i-Mode data is transmitted from each of the fixed radio ports that received the data together with the address of the moving base station to which the data is directed.

The transmission of i-Mode data from different fixed radio ports is intentionally staggered by introducing different transmission delays so that the signals can be received and separated at the moving base stations. The transmission delays can be precisely controlled by means of synchronous distribution via the optical ring in SONET or SDH format. The processor in the receiving moving base station compares, aligns, and combines the multiple copies of the received i-Mode data signals for the best reception.

THE MOBILE UNIT-MOVING BASE STATION INTERFACE

The communication interface between a moving base station and the mobile units is direct sequence spread spectrum CDMA. Each moving base station communicates with a maximum of 12 mobile units through 12 20-Mbps channels. Preferably, at least one channel is reserved to accommodate mobile unit handoffs. Although the overall bit rate is 25 Mbps, 25 percent of the channel is used for error correction, resulting in the 20-Mbps effective i-Mode data rate. Due to the application of the spreading function, each 25-Mbps channel is spread to the 2500-MHz bandwidth to achieve a processing gain of approximately 18 to 20 dB. This is sufficient for the short-range line-of-sight radio signal propagation.

TRANSMISSION LAYERS

As shown in the system interconnection model in Figure 13-3, the moving base station mobility is concealed within the lower transmission layers and transparent to the upper layers.[6] Accordingly, functions, such as registration, authentication, and paging as well as control, signaling, and traffic channels, are implemented at a transmission layer transparent to the fixed radio ports and the gateway. The mobility of the moving base stations is supported by the functions and subsystems illustrated in Figure 13-3.

SYSTEM CAPABILITIES

The proposed high i-Mode user capacity infrastructure provides 20-Mbps i-Mode data channels to mobile units traveling at speeds anywhere from 0 to over 100 mph. I-Mode users traveling on highways can productively utilize time spent in a vehicle

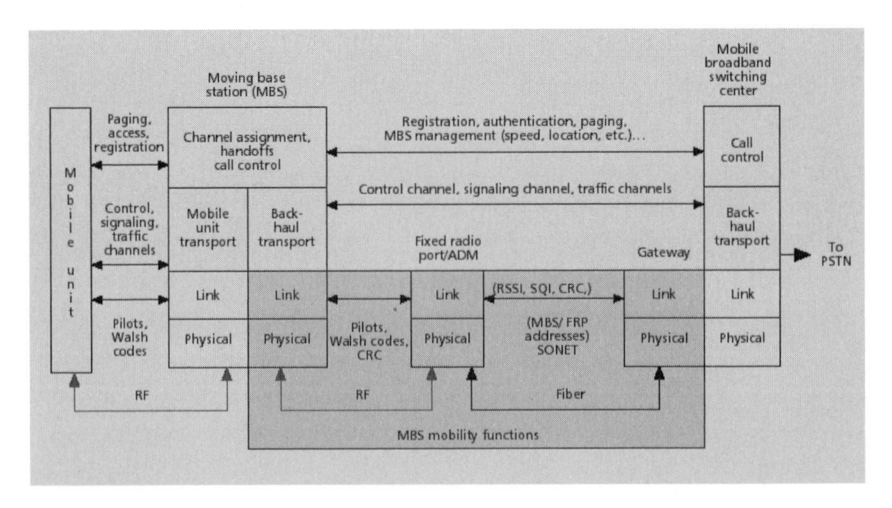

FIGURE 13-3 A system interconnection model for a moving base station communication system.

or enjoy the various entertainment or convenience applications only achievable using high-i-Mode data-rate channels. As described next, the cost per channel of the proposed system is comparable to the cost per channel of wireline networks.

COMPARISON TO OTHER SYSTEMS

At first glance, some existing communication systems appear to provide similar communication services as the proposed system. After careful examination, however, it is apparent that these other systems are limited in i-Mode user capacity, bandwidth, and/or user speed. Figure 13-4 illustrates a comparison between mobility and i-Mode data rates for the moving base station system and other proposed and existing communication systems. [7]

As explained in the preceding, systems using only large cells are limited in capacity and bandwidth. Due to finite available frequency spectrum, the size of a cell and the i-Mode data rate of a channel dictate the capacity of the system. By reducing the size of the cell, capacity, and channel, bandwidth is increased at the cost of i-Mode user speed.

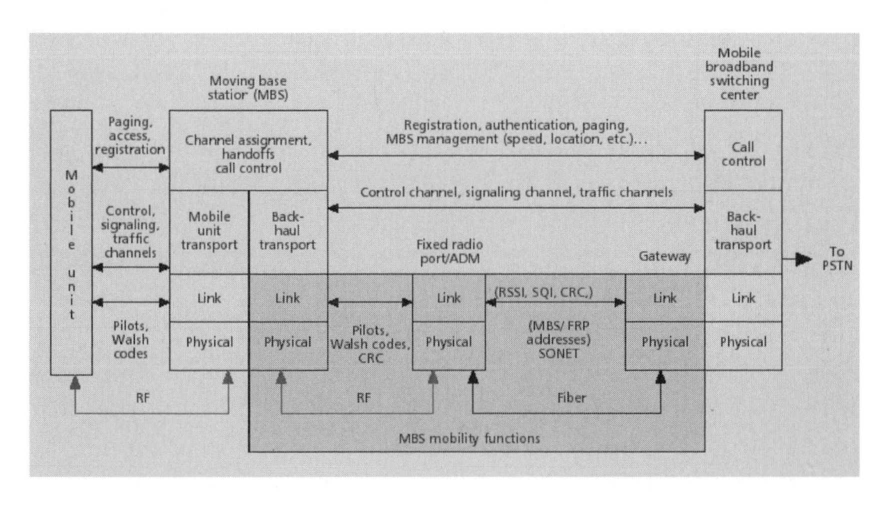

FIGURE 13-4 A comparison of moving base station systems to other proposed and existing infrastructures.

One approach taking advantage of both small-cell and large-cell systems includes using two cell sizes where a large cell and several small cells overlap. The use of a large cell, however, still limits the bandwidth and number of i-Mode users traveling at high speeds. Such a two-tier system is severely limited in areas where a high concentration of i-Mode users is traveling at high speed. For example, the 3G proposal that takes advantage of the two-tier approach, still limits high-speed i-Mode users to only 144-Kbps channels, although the proposed moving base station system can provide channels on the order of 20 Mbps.

Although satellite systems appear to provide huge i-Mode data channels, after careful examination, it is clear that proposed satellite systems are not well suited for providing broadband channels to high concentrations of i-Mode users traveling at highway speeds due to the fact that extremely large cells are used.

System Cost and Revenue Overview

The wireless infrastructure portion of the proposed highway broadband system is expected to cost less than $1.9 million/mile of roadway. This wireless infrastructure cost estimate is based on costs of existing infrastructures and pricing quotations from several sources and includes costs for the two main rails, the center rail, 52 active moving base stations, 13 spare moving base stations, and 26 fixed radio ports without ADMs. As in all wireless infrastructures, additional costs will apply for the backhaul, which will include elements such as fiber optic cable, ADMs, and gateways.

The total infrastructure cost for the proposed system, including the radio portion and backhaul to the MBSC, is less than $4.3 million/mile. The division of costs associated with the radio portion and backhaul is provided to emphasize that most of the system cost is associated with the backhaul and that any system providing such a large number of 20-Mbps channels in a highly concentrated area will most likely require a similar sophisticated backhaul infrastructure.

SYSTEM COST AND REVENUE ANALYSIS

Although additional investigation is required to obtain estimates with a high level of confidence, the following cost analysis is provided as a starting point for future discussions and research. At the request of an information source, some of the estimates used for the analysis are not referenced in order to preserve the confidentiality of proprietary discussions between the source and third parties. Cost and revenue estimates are presented in Table 13-1 and Table 13-2, respectively.[8]

TABLE 13-1 Infrastructure Cost Per Mile

EQUIPMENT	COST/UNIT	QTY/MILE	COST/MILE
Conveyor System rails (two main plus one center)	N/A	N/A	$565,000
Moving base station carriage (main rails)	$2,500	52	$130,000
Moving base station carriage (for center rail and spares)	$2,500	13	$32,500
Moving base station circuitry (main rails)	$10,000	52	$520,000
Moving base station circuitry (for center rail and spares)	$10,000	13	$130,000
Radio port-ADM	$85,000	26	$2,210,000
Radio port-radio circuits	$10,000	26	$260,000
Radio port-other	$10,000	26	$260,000
Fiber optic cable	N/A	N/A	$100,000
Gateway (including ADM)	$500,000	0.08 (2 per 25 miles)	$40,000
Total	N/A	N/A	$4,247,500

TABLE 13-2 Revenue Per Year Per Mile

Parameter	Assumption	Assumption	Calculation	Result
Average number of vehicles per 200 foot lane	35 feet spacing	N/A	200 ÷ 35	6
Average number of vehicles per moving base station	3 lanes/highway	200 feet spacing between moving base stations	6 (3)	18
Erlangs of communic-cation traffic	50 percent usage/vehicle	N/A	18 (50/100)	9
Channel occupancy	50 percent usage/vehicle	12 total traffic channels per moving base station	(9 ÷ 12)(100)	75 percent
Number of base stations per mile	200 feet spacing between moving base stations	Two-way highway	(5280 ÷ 200) (2)	52
Average number of channels in use per moving base station	50 percent usage/vehicle	N/A	18 (50/100)	
Average number of channels in use per mile	52 moving base stations/ mile	9 channels/ moving base station	52 (9)	468
Revenue/ minute per mile	$0.50/min/ channel	N/A	0.50 (468)	$234
Number of revenue minutes per year	8 hours rush hour/day	250 rush hour days per year	8 (60) (250)	120,000

A monorail transportation design expert estimated that a suitable conveyor system would likely cost about $480,000 for the two main rails along both sides of a mile-long stretch of highway. Based on the information provided by the expert, it is estimated that each additional rail implemented adjacent to one of the main rails will cost about $85,000. The cost of the conveyor infrastructure therefore including the center rail is $565,000.

The carriage portion for each moving base station was estimated at $2,500 by the monorail expert. The carriage includes the chassis in addition to all motors, control systems, power systems, and mechanical components.

The analog and digital circuitry performing the communication functions of the moving base stations, including power circuitry and antennas, is expected to cost approximately $10,000. This estimate is based on high-volume costs for existing small-cell systems such as the *Personal Handiphone System* (PHS). Development costs associated with the base stations will most likely be associated with the 60-GHz aspect of the design, because the remainder of the required circuitry is similar to current cellular and *personal communications services* (PCS) designs.

Although initial development costs may require additional expenditure to cover issues likely to be encountered with 60-GHz designs, these costs are likely to be recouped in the high-volume manufacturing advantages, size, and integration benefits of high-frequency circuitry. Technology currently being developed for other 60-GHz applications may provide shortcuts for much of the required development. For example, base stations for fixed applications are currently commercially available and may be modified to accomplish the desired RF functions of the proposed system. It is clear, however, that further research and development is required to produce the 60-GHz circuitry included in the moving base station.

The majority of the cost of a fixed radio port is due to the high-speed ADMs needed to couple the signals to the fiber optic cable. Based on an estimate obtained from a manufacturer, an ADM should cost approximately $85,000 in high volume. Due to the similarities in requirements, the RF circuitry portion of the radio port is estimated to cost the same as the moving base station RF circuitry. An additional $10,000 is included to cover miscellaneous costs, such as those associated with building an adequate structure for supporting the radio portion of the fixed radio port. A fixed radio port, including an ADM, radio circuitry, and tower, is estimated to cost no more than $105,000 in high volume.

The cost of implementing fiber optic cable appears to be continually falling. At least one source estimates the cost for implementing fiber along a highway to be $100,000/mile.

In the proposed system, a single gateway for each 25-mile section of roadway, should be sufficient for performing the required functions. In order to increase the robustness of the system and minimize potential service interruption, a redundant gateway is included for each 25-mile section of highway.

Therefore, the total cost of the moving base station infrastructure, including backhaul, is approximately $4.24 million/mile. As mentioned in the preceding, the various cost estimates were either obtained from manufacturers, based on similar existing systems, or derived using a combination of information resources. Additional research is required to obtain manufacturing cost estimates with a higher level of precision.

Revenue

Revenue generation is, of course, highly dependent on the demand for wireless services, and further market research is needed to obtain predictions of consumer needs for the future. Accordingly, the following assumptions may be interpreted as being either too aggressive or too conservative due to the variance of predictions related to consumer demands for wireless service. The overriding assumption is that consumers will use

20-Mbps channels while on the highway if the cost per minute is comparable to other landline and wireless services.

The estimates are based on a typical three-lane urban highway that experiences rush hour traffic for four hours in the morning and four hours in the afternoon. In order to simplify the revenue estimates, only rush hour traffic is included in the estimates. This simplification appears reasonable because these hours will experience the highest amount of traffic, and include the highest numbers of wireless and Internet users with a desire to minimize downtime while driving to work. Additional revenue will likely be generated during the weekends and no-rush hour times, but is omitted from the revenue estimate.

Health Concerns

Due to the high frequencies utilized by the proposed system, concerns regarding human exposure to RF signals are likely to be raised. It is clear that all electromagnetic radiation should be continually examined and that the effects of RF signals on the human body should continually be researched.

The small-cell infrastructure enables the transmission of signals at power levels on the order of 1 to 10 milliwatts (mW), resulting in an overall lower level of RF radiation exposure than with other systems. Furthermore, higher-frequency signals have lower penetration into the human body.

CONCLUSION

This chapter discussed 3G broadband mobile communications. In conclusion, the proposed wireless system previously discussed may provide a unique solution for the highway portion of the high-bandwidth global communication system of the future. Utilizing many proven technologies in conjunction with a moving cell infrastructure results in a relatively inexpensive approach to providing high-bandwidth communication channels to a large number of i-Mode users traveling at highway speeds. The scene of base stations gliding along the centers of the highways of tomorrow may soon be a sight as common as telephone poles lining the highways of today.

END NOTES

[1]John R. Vacca, *Wireless Broadband Networks Handbook: 3G, LMDS, and Wireless Internet*, McGraw-Hill, New York, 2001.

[2]"NTT DoCoMo Finally Delivers 3G Videophones," AnywhereYouGo.com, 3000 Waterview Parkway, B2E14, Richardson, TX 75080, 2001.

[3]John R. Vacca, *The Cabling Handbook* (2nd Edition), Prentice Hall, Upper Saddle River, N.J., 2001.

[4]Charles D. Gavrilovich, Jr. and Cary Ware & Freidenrich, LLP, "*Broadband Communication On The Highways Of Tomorrow*," Gray Cary & Freidenrich LLP, 139 Townsend Street, Suite 400, P.O. Box 77630, San Francisco, CA 94107, 2001.

[5]Ibid.

[6]Ibid.

[7]Ibid.

[8]Ibid.

PART IV

CONFIGURING I-MODE

CHAPTER FOURTEEN

I-MODE AND JAVA

NTT DoCoMo, in cooperation with Sun Microsystems (USA), Sun Microsystems K.K. (Japan), and handset manufacturers, has developed a Java programming environment for i-Mode handsets. Depending on how it is utilized, i-Mode content developers can create content for users that is much more dynamic and interactive than what has been possible until now with HTML-based content. This chapter offers an explanation of the i-Mode Java environment and language specifications that are utilized in 50X series devices.

WARNING

Please be aware that the Java for i-Mode specifications mentioned in this chapter may change without prior notice. Please be also aware that this author, McGraw-Hill, and NTT DoCoMo accept no responsibility for said changes.

WHAT ABOUT I-MODE AND JAVA?

Java-enabled handsets recently have been introduced in Japan. The Java-enabled i-Mode handsets (series 503i) went on sale recently and are now generally available in Japan. Java-enabled handsets contain the *K Virtual Machine* (KVM) from SUN—separate from the browser. Users are able to download Java applets into the handsets for games, agent-type services, and other applications.

I-Mode Java Specifications: I-Mode Java Main Features

The Java platform was developed specifically for 50X series mobile devices. I-Mode Java is based on the following (see Figure 14-1)[1]: Sun Microsystem's *Java™ 2 Platform Micro Edition Connected Limited Device Configuration* (J2ME CLDC) targeted at consumer electronics and embedded devices (see sidebars, "Java™ 2 Platform, Micro Edition (J2ME™ Platform)" and "Java 2 Platform, Micro Edition"), and the i-Mode extension library (i-Mode Java profile) defining the user interface and *Hypertext Transfer Protocol* (HTTP) transfer, and so on. The i-Mode extension library will be made available in the near future.

All of the i-Mode network architecture after the introduction of Java for i-Mode will remain the same as the current i-Mode architecture. Java programs from Web sites will download together with normal HTML documents through the same HTTP protocol, and Java programs that run internally from the mobile phone handset will also communicate with the internet through HTTP (see Figure 14-2).[4]

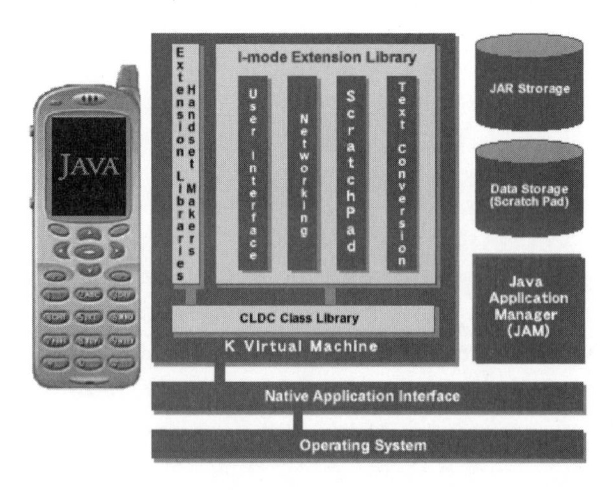

FIGURE 14-1 The i-mode Java-enabled handset's internal architecture.

JAVA™ 2 PLATFORM, MICRO EDITION (J2ME™ PLATFORM)

Recognizing that *one size doesn't fit all*, Sun has regrouped its innovative Java™ technologies into three editions: Micro (J2ME™ technology), Standard (J2SE™ technology), and Enterprise (J2EE™ technology). Each edition is a developer treasure chest of tools and supplies that can be used with a particular product:

- Java virtual machines that fit inside the range of consumer devices
- A library of APIs that are specialized for each type of device
- Tools for deployment and device configuration
- A profile, that is, a specification of the minimum set of APIs useful for a particular kind of consumer device (set-top, screenphone, wireless, car, and digital assistant) and a specification of the Java virtual machine functions required to support those APIs

J2ME technology specifically addresses the vast consumer space, which covers the range of extremely tiny commodities, such as smart cards or a pager all the way up to the set-top box, an appliance almost as powerful as a computer. Like the other editions, the J2ME platform maintains the qualities that Java technology has become famous for:

- Built-in consistency across products in terms of running anywhere, any time, over any device.
- Portability of the code.
- Leveraging of the same Java programming language.
- Safe network delivery.
- Applications written with J2ME technology are upwardly scalable to work with the J2SE and J2EE platforms.

With the delivery of the J2ME platform, Sun provides a complete, end-to-end solution for creating state-of-the-art networked products and applications for the consumer and embedded market. J2ME technology enables device manufacturers, service providers, and content creators to gain a competitive advantage and capitalize on new revenue streams by rapidly and cost-effectively developing and deploying compelling new applications and services to their customers worldwide.[2]

JAVA 2 PLATFORM, MICRO EDITION

The CLDC is a Java community process effort (JSR-30) that has standardized a portable, minimum-footprint Java™ building block for small, resource-constrained devices. The CLDC configuration of J2ME provides for a virtual machine and a set of core libraries appropriate for use within an industry-defined profile, such as the wireless profile as specified by the *Mobile Information Device* (MID) specification (JSR-37).

The CLDC standardization effort is the result of a collaboration amongst 18 companies representing several different industries. Target devices for CLDC are characterized generally as follows:

- 160KB to 512KB of total memory, including both RAM and flash or ROM, available for the Java platform
- Limited power, often battery-powered operation
- Connectivity to some kind of network, often with a wireless, intermittent connection and with limited (often 9600 bps or less) bandwidth
- User interfaces with varying degrees of sophistication or even none

Cell phones, two-way pagers, *personal digital assistants* (PDAs), organizers, home appliances, point-of-sale terminals, and car navigation systems are some of the devices that might be supported by CLDC. The CLDC reference implementation utilizes Sun's KVM and is the first Sun product to contain a commercial use-ready implementation of the KVM.

As previously explained, Java technology-based i-Mode services are now available. This new service is called i appli.

I APPLI

Recently, Sun Microsystems Inc. applauded NTT DoCoMo Inc.'s (NTT DoCoMo) launch of Java™ technology-enabled i-Mode phones and services in Japan. The new i-Mode service,

IMPROVEMENTS IN VERSION 1.0.2

The latest release of J2ME CLDC contains an updated version of the KVM, and its code has been largely rewritten to bring significant improvements:

- Faster bytecode interpreter
- Exact, compacting garbage collector
- Java-level debugging APIs (based on the existing JDWP standard)
- Preverifier improvements
- Large number of bug fixes and smaller enhancements

This release also features the first implementation of J2ME CLDC for Linux.

IMPORTANT INFORMATION

The J2ME CLDC reference implementation is a source code product that is provided for porting to various platforms. In order to ship this product commercially, a commercial use agreement must be entered into with Sun. As part of any commercial use agreement, it is necessary to pass compatibility tests that ensure that we all benefit from a standardized Java application environment. The *technology compatibility toolkit* (TCK) for CLDC can be licensed from Sun separately.[3]

called i appli, opens the next chapter of i-Mode revolution by providing consumers with new interactive content including multi-player games, mobile e-commerce,[5] and, in the future, enhanced communications.

NOTE

I appli is DoCoMo's trade name (trademark) for Java applications that can be downloaded via DoCoMo's

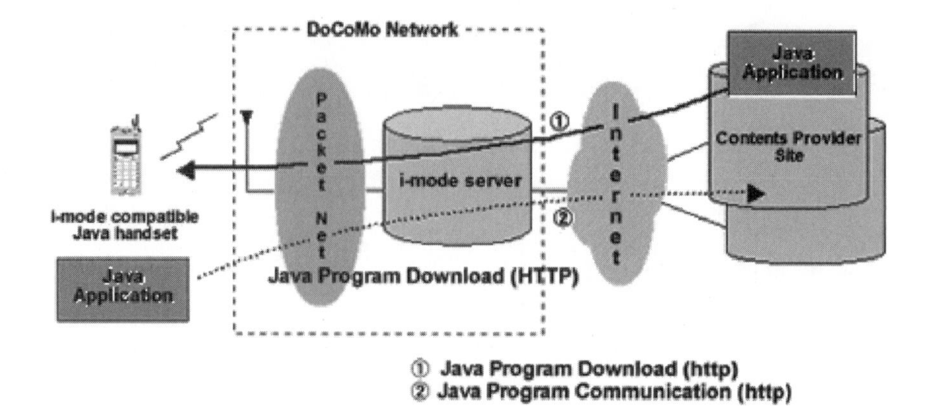

FIGURE 14-2 Overall i-Mode network architecture.

i-Mode mobile Internet service to Java enabled i-Mode handsets. As of this writing, approximately 44 official DoCoMo partner sites are offering i appli Java applets for download.

As previously explained numerous times, NTT DoCoMo is Japan's number one mobile phone company and has had phenomenal success to date with its i-Mode service, signing up more than 31 million subscribers since the service was introduced in February 1999. Now with the launch of Java technology-enabled phones and services, NTT DoCoMo can offer highly differentiated products and services that provide consumers dynamic, personalized, and interactive content.

NTT DoCoMo's launch of its i appli services is a significant event for both Sun and the Java platform. Through Java technology, NTT DoCoMo brings advanced capabilities to the wireless services marketplace that will enable consumers to take advantage of the many innovative new services being created in Java technology for mobile handsets.

NOTE

Java-enabled handsets are on sale by DoCoMo in Japan. The Java-enabled handsets contain the KVM from Sun.

At present (as of this writing), two types of handsets are available: P503i and F503i.

Java technology advances the extraordinary success of i-Mode by improving the capabilities of the next generation of i-Mode handsets and services for customers. Sun's Java technologies and world-class carrier-grade services provide the ideal foundation for NTT DoCoMo to build and expand on their leadership in the wireless market.

The launch of Java technology-based i-Mode services represents an important milestone for Java technology. The new 503i cellular phones from Fujitsu and Matsushita (Panasonic) are among the first to be enabled with the J2ME technology. Additional i appli handsets incorporating J2ME technology will be provided by other manufacturers in 2002. J2ME technology-enabled interactive services are the next step beyond today's text-based static content. Java software enhances the user experience by supporting easy-to-use, graphical, interactive services for wireless devices. With J2ME technology-enabled phones, users can download software to handsets for disconnected use, as well as access applications interactively from the network.

Initial i appli applications include a mix of computer games and financial services. Over time, the i appli service will offer new animated games with enhanced graphics and high-fidelity sound, software for chatrooms, location-based services with zooming maps, secure mobile commerce, and business support programs such as groupware. Already, numerous high-profile content partners have signed up to create new applications and services based on Java technology including Disney, Bandai, Namco, Sega, and DLJ.

CONCLUSION

This chapter discussed i-Mode and Java. In conclusion, NTT DoCoMo recently entered into joint ventures to significantly expand i-Mode's presence in Europe and to support Java-based wireless services in Japan.

The Japanese wireless company, which recently took a 16- percent stake in AT&T Wireless in the United States, indicated the joint venture with Holland's KPN Mobile and Italy's *Telecom Italia Mobile* (TIM) will result in a pan-European i-Mode portal. The companies indicated that they will roll out the i-Mode services in Europe later in 2002. Initially, the service will target KPN's and TIM's customers in Holland, Italy, Germany, and Belgium.

In addition, DoCoMo and KPN have entered a memorandum of understanding to create a new entity that will consolidate the wireless data businesses of KPN Mobile in the Netherlands, KPN Orange in Belgium, and E-Plus in Germany. DoCoMo will own 25 percent of the new unit.

Separately, as previously explained, DoCoMo launched new i-Mode services based on Java technology. The new service is now available throughout Japan. Thirty-eight special sites are now available exclusively to users of Java-enabled wireless phones, which DoCoMo sells.

DoCoMo also indicated that applications can be downloaded to the phones so a constant connection won't be necessary. Finally, DoCoMo expected the phones to be used for business applications, such as mobile e-commerce, as well as for entertainment, such as games and karaoke.

END NOTES

[1]WestCyber Corporation, 1350 Broadway, Rm. 705, New York, NY 10123, 2001.

[2]Sun Microsystems, Inc., 901 San Antonio Road, Palo Alto, CA 94303 USA, 2001.

[3]Ibid.

[4]WestCyber Corporation, 1350 Broadway, Rm. 705, New York, NY 10123, 2001.

[5]John R. Vacca, *Electronic Commerce: Online Ordering and Digital Money with CD-ROM*, Charles River Media, 2001.

CHAPTER FIFTEEN

I-MODE HANDSETS

Why is i-Mode generating such intense industry buzz? Why is DoCoMo being courted with such reverent ardor by heavyweight Net content providers, platform owners, and mobile operators in North America, Asia, and Europe? Recently, AOL, Yahoo, Palm Computing KK, U.S. telcos BellSouth and SBC Communications, Hong Kong's Hutchison, the United Kingdom's Hutchison 3G UK Holdings (proud owner of a spanking new 3G license), and Holland's KPN Mobile have all shared tie-up or capital investment headlines with DoCoMo. Invariably, i-Mode has been cited as the Japanese partner's prime asset. Why? Is i-Mode some stellar new example of handset engineering wizardry that gives DoCoMo a proprietary and unique technological lock on the mobile Internet?

The answer is a resounding yes! I-Mode handset technology is the secret to its success.

PRESENT AND FUTURE HANDSETS FOR USE WITH THE I-MODE SYSTEM

Certainly, there is a lot to admire about the i-Mode handset's success. Thirty-two million-plus subscribers, 26,000-plus official and nonofficial content partners, massive revenues from data packet transmissions, a healthy boost to voice traffic, and all those mobile handset surfers calling in to make ticket or video rental reservations (after confirming availability via the small handset screen)—all these help validate i-Mode handset technology as a pretty impressive achievement. NTT DoCoMo is, by the way, now the world's largest wireless ISP.[1] Table 15-1 shows the latest NTT-DoCoMo i-Mode handsets (discontinued and non-i-Mode handsets are not included).[2]

Letters designate manufacturers: *SO* = Sony and *N* = NEC. Thus, *SO503i* is the 503i-s handset by Sony, and N503i is the 503J-series handset by NEC and so on.

Table 15-2 shows handsets for the 3G/FOMA test phase.[3] These handsets and PC-cards are at the moment in testing among approximately 4,000 test users in Japan.

Basically, the i-Mode handset has taken off because

- It is attractive to content providers because it manages their paid services for them (and DoCoMo isn't being too greedy about doing so).
- The phones have a cool little *i* button to give the user direct access to the content sites—there's no hint of the Internet nor any of its complications.
- The content really is good.
- I-Mode chose to use *compact HTML* (cHTML) while *Wireless Application Protocol's* (WAP's) *Wireless Markup Language* (WML) was still in its infancy, which made it really easy for content providers and users to create sites.
- In this land of great public transportation and long lines, i-Mode gave its subscribers something cool to do while waiting or commuting.[4]

CONCLUSION

This chapter very briefly discussed i-Mode handsets. In conclusion, nothing is really magical about it at all, with no proprietary handset technology, and nothing that any other network operator couldn't replicate. Other operators don't really seem to be trying though. Ask anyone creating a mobile Net service here in Japan—they're going to i-Mode first.

I-Mode handsets are an extraordinarily clever business, marketing, and service hardware offering, targeted at two superdense population centers (Tokyo and Osaka) in a country that suffered (and suffers) from the world's highest-cost wired Internet access. Initially, a consulting company worked the initial i-Mode service model and didn't get much credit for doing so, but did manage to bring up and resolve the following issues:

- The debate over the two fee options (tenant versus pay-per-packet)
- Why compact HTML won over WAP's WML
- How the minimum screen size (eight kanji characters—since increased) and maximum handset weight (100 grams) were decided
- How and why banks were targeted for initial inclusion as content partners
- Why only seven journalists attended the first i-Mode news conference, but 500-plus attended the second [5]

However, the question remains: Why are the international partners lining up to write i-Mode handsets onto their dance card? Why not just adopt the i-Mode model, and leave DoCoMo out of it?

TABLE 15-1 The Latest NTT-DoCoMo I-Mode Handsets

Handset	Weight	Battery (standby)	Battery (talking)	Screen color	Comments
SOS03i	115 g	210 hours	140 minutes	65,536 colors (TFT)	Java (i appli)
N503i	98 g	. . . hours	. . . minutes	4096 colors	Java (i appli)
F503i	77 g	430 hours	135 minutes	256 colors	Java (i appli)
P503i	74 g	400 hours	140 minutes	256 colors	Java (i appli)
R691i (Geofre)	99 g	430 hours	120 minutes	Gray scale	water-resistant
P209iS	84 g	380 hours	135 minutes	256 colors	. . .
D209i	74 g	400 hours	120 minutes	256 colors	. . .
P209i	60 g	350 hours	135 minutes	Gray scale (4)	. . .
N209i	86 g	500 hours	120 minutes	Gray scale (4)	. . .
F209i	63 g	450 hours	135 minutes	256 colors	. . .
R209i	63 g	430 hours	120 minutes	Gray scale	. . .
KO209i	69 g	350 hours	120 minutes	256 colors	. . .
ER209i	77 g	310 hours	130 minutes	B&W	. . .
N502it	105 g	460 hours	130 minutes	256 colors	. . .
P502i	69 g	300 hours	125 minutes	Gray scale (4)	. . .
N502i	98 g	420 hours	120 minutes	Gray scale (4)	. . .

Handset	Weight	Battery (standby)	Battery (talking)	Screen color	Comments
D502i	84 g	350 hours	130 minutes	256 colors	. . .
SO502i	73 g	210 hours	120 minutes	Gray scale	. . .
SO502iWM	120 g	200 hours	100 minutes	256 colors	Walkman type
NMS02i	77 g	270 hours	130 minutes	B&W	(Discontinued)
P821i	82 g	210 to 340 hours	115 minutes to 6.5 hours	Gray scale (4)	Doccimo
N821i	105 g	290 to 600 hours	120 minutes to 7 hours	Gray scale (4)	Doccimo
SH821i	76 g	220 to 400 hours	110 minutes to 7.5 hours	256 colors	Doccimo

TABLE 15-2 Handsets for the 3G/FOMA Test Phase

Handset	Weight	Battery (standby)	Battery (talking)	Description	Comments
FOMA N2001	105 g	. . .	. . .	Color screen, multitasking	Simultaneous i-Mode and voice
FOMA P2101V	150 g	. . .	. . .	Color screen, video camera	Two-way video
FOMA P2401	50 g	. . .	. . .	PC-card (for data transmission from PC)	Downlink: 384 Kbps, Uplink: 64 Kbps

It's safe to suppose that the mobile operators, for their part, are looking to DoCoMo to show them how to do it—the business, that is, not the handset technology. Palm Computing is talking to DoCoMo for the mundane reason that it wants to sell wireless Palm VII-style handhelds to the gadget-crazy Japanese, and DoCoMo is by far and away the domestic heavyweight. There's no need to even talk to the other wireless network operators until after an i-Mode deal is sewn up.

In the final analysis, both AOL and Yahoo are probably looking to partner with DoCoMo (see sidebar, "English Content Pushes I-Mode Beyond Japan") more for the bragging rights than for access to the 32 million i-Mode handset customers, most of whom aren't surfing, but talking. In the battle for eyeballs back home, the bigger the numbers of the international partners, the happier the corporate hype-meisters.

ENGLISH CONTENT PUSHES I-MODE BEYOND JAPAN

Japan's largest cellular telephone carrier, NTT DoCoMo, and AOL are close to a deal that will enable customers of cellular Internet services to access AOL content on their telephone handsets. This is the latest of several partnerships between content providers and telecommunications giants, but the first to focus on mobile Internet users.

If, as expected, the deal goes ahead, AOL content will be offered, not just to the Japanese market, where NTT DoCoMo has 32 million users on its i-Mode wireless Internet service, but to other areas of the world as well. When asked if the service might extend to the Middle East, a spokesman for AOL declined to comment at this time.

I-MODE TECHNOLOGY

NTT DoCoMo has made clear its ambition to push i-Mode technology overseas. WAP-based systems are currently in operation on a global scale, most recently deployed in the Middle East.

However, few platforms have matched the popularity of i-Mode, and NTT DoCoMo is keen to extend its influence in the wireless Internet sector. A joint venture with AOL would offer new English content to mobile users and drive the adoption of i-Mode technology.[6]

END NOTES

[1]John R. Vacca, *Wireless Broadband Networks Handbook: 3G, LMDS, and Wireless Internet*, McGraw-Hill, New York, 2001.

[2]Eurotechnology Japan K.K., Parkwest Building 11th Floor, 6-12-1 Nishi-Shinjuku, Shinjuku-ku, Tokyo 160-0023, Japan, 2001.

[3]Ibid.

[4] Daniel Scuka, *What's So Great About I-Mode?* LINC Media Inc., 1745 S. Alma School Rd., STE 245, Mesa, AZ 85210-3013, USA, 2001.

[5]Ibid.

[6]The Information & Technology Publishing Co. Ltd., P.O. Box 500024, Dubai, United Arab Emirates, 2001.

PART V

MANAGING I-MODE MARKETS

CHAPTER SIXTEEN

OVERVIEW OF I-MODE MARKETS

Most attempts to interpret the rapidly shifting world of telecommunications make the mistake of viewing the balance of power as bipolar. The North Americans pretty much invented the Internet and so are better at fixed-line communications than anyone else. The likes of Lucent, Cisco, and Nortel dominate as manufacturers, just as operators, such as WorldCom or Global Crossing, export fixed-line service expertise.

On the other hand, the Europeans have carved out a reputation as masters of the wireless world. Thus, higher mobile phone penetration rates have helped Nokia and Ericsson climb to the top of the markets for mobile handsets and infrastructure. Vodafone remains the largest international operator.

There are plenty of exceptions, of course. Motorola, the U.S. cellular giant, has recovered from a crisis of confidence that followed the collapse of its satellite phone venture Iridium to once again challenge Nokia for leadership of the world handset market. European manufacturers, such as Alcatel, Siemens, and Marconi, are no slouches when it comes to wires and switches.

Nevertheless, the perception persists that America still does not get the wireless Internet while Europe's protected phone monopolies are preventing it from wiring up for the real PC revolution. Apart from the crudeness of such analysis, it ignores Asia, the fastest growing telecoms market in the world.

In recent years, Asian telecoms had looked rather like a sleeping giant. The richest market, Japan, was highly advanced, but its decision to pursue a separate technology path for mobile phones prevented both operators and manufacturers from realizing their potential abroad. NTT sheltered under the protection of some of the highest fixed line tariffs in the developed world.

Yet, recently, there have been some dramatic shifts. NTT DoCoMo, NTT's mobile phone arm, has established an early head start in the race to encourage customers to use their handsets for more than just voice calls.

DoCoMo's packet-switched i-Mode technology proves that the wireless Internet is more than just a pipe dream and shows what can be achieved when operators move to similar packet-switched systems (which unlike circuit-switched networks, offer an *always on* capability) in Europe and the United States.

More importantly, DoCoMo has demonstrated the first signs of its ambition to exploit this advantage abroad. It has taken minority stakes in British and Dutch operators and hopes to show the Europeans some of the joys of i-Mode before long. Even DoCoMo's overly cautious parent has finally taken the plunge with a controversial bid for Verio, a U.S. Internet company, despite criticism that NTT's domestic market remains over protected.

Western eyes have also turned to Hong Kong, where Li Ka-shing's Hutchison Whampoa has proved that conglomerates can be nimble too. Hutchison's profitable sale of Orange in the United Kingdom may have led many to conclude it was a trader rather than an operator, but the subsequent reentry into the British market for a far lower price was one of the great surprises of the *third generation* (3G) mobile phone auctions.

In the world of fixed-line Internet, the Li family has been busy too. Richard Li's audacious takeover of Hong Kong Telecom, using the over-inflated shares of Pacific Century CyberWorks, arguably marked the high point (or low point) of the dotcom investment bubble in a far more dramatic way than the merger of America Online and Time Warner.

Yet, the potential of both Japan and Hong Kong to capture the imagination of investors is dwarfed by their giant neighbor, China. It is here, in a country of 1.3 billion people, that the scale of the worldwide revolution in communications becomes apparent.

When China's ministry of information wanted to improve communication links to the far-flung rural provinces of western China and Tibet, it could not afford to hire thousands of mechanical diggers to lay the enormously long fiber optic cable required.[1] However, it could turn to the Peoples Liberation Army and use thousands of troops to dig trenches over the Himalayas by hand in an act of human endurance that is reminiscent of the construction of the Great Wall.

NOTE

China now buys more mobile telecoms equipment from multinational suppliers than any market outside the United States.

Similar heroic efforts are taking place in India, where the government is encouraging companies to lay fiber optic cables alongside the vast railway network—reusing older communication routes in just the same way as the canals, motorways, and electricity grids have helped string uninterrupted fiber across Europe. However, size is not everything. What differentiates China is the relative wealth of its teeming masses.

In the eastern cities at least, Chinese middle classes do not just offer the potential to become the world's most valuable telecoms market—they already are. One mobile phone operator alone, China Mobile, is adding around 2 million new subscribers a month and should shortly overtake Vodafone as the world's biggest operator before it even ventures abroad.

Overall penetration rates remain tiny, but the 59 million mobile subscribers across China today buy more equipment from multinational suppliers, such as Ericsson, Motorola, and Nokia, than any market outside the United States. All three are investing at breakneck speed in the belief that China will not be number two for long.

Technology is catching up, too. By requiring Western manufacturers to enter into joint ventures with local companies and build factories in China, the communist government has skillfully copied many of the high-tech electronics essential to modern telecoms infrastructure.

Returning Chinese émigrés from America, Hong Kong, and Taiwan are joining others who were educated abroad to create a vibrant entrepreneurial culture ready to exploit this know-how of how to build innovative technology companies of their own. Regardless of their public confidence in Chinese government promises to keep the door open, this is a serious threat to the supremacy of European and U.S. manufacturers.

OPPORTUNITIES

Even if Chinese technology, such as the 3G mobile phone standard *Time Division-System Code Division Multiple Access* (TD-SCDMA), remains less than cutting edge, adoption by giant operators, such as China Mobile, will always guarantee it a place on the world stage. The opportunities for U.S. and European operators to break into the long-protected Chinese market also look limited. Despite China's accession to the World Trade Organization, the government retains the power to prevent foreigners from taking majority control of any domestic phone operator.

When even Hong Kong operators, such as the enviably well-connected Mr. Ka-shing are still treated as outsiders, it is hard to see what chance Vodafone or AT&T have to make a serious impact on the mainland. Government attempts to control Internet content for political reasons will also hamper the growth of foreign content providers—as one U.S. Web site recently found to its cost when it accidentally referred to Taiwan as an independent country and was rumored to be blocked for months in retaliation.

Censorship of the Internet may also hold back China's economic growth in the short term, but enough Internet-savvy managers are at the top of the industry to ensure that Chinese politics does not interfere with the money-making too much. All of this takes money, of course, and the future of companies, such as China Mobile and China Netcom, depends on their ability to raise investment funds on the international capital markets. So far, investors in Hong Kong and New York have fallen over themselves to throw money at China, despite primitive corporate governance and limited transparency.

In future, the enormous and simultaneous demand for capital from the globalized telecoms industry could hinder those who are not prepared to play by the rules of the global markets. Yet, Chinese operators are in a no worse position than many other cash-strapped companies anxious to secure funding before the tap runs dry. If only the biggest survive the shakedown, then China cannot be ignored for long.

OVERVIEW OF THE JAPANESE CELLULAR MARKETS

Japan's cellular market is the second largest in the world and is about to be opened and deregulated. This produces a large number of new business opportunities for foreign corporations in Japan.

Japan's *Telecommunications Advisory Council* (TAC) recently prepared far-reaching recommendations for the Japanese government to support the growth of the cellular and multimedia markets. TAC estimates that until the year 2010, building the info-communications infrastructure will create 3,540,000 new jobs and will lead to a new demand on the order of $600 billion for products and services related to fiber optics and demand for an additional $700 billion for conventional products, such as video recorders, telecommunications equipment, computers, and software.

Both the demand for cellular services, as well as the investments by Japan's telecommunication service providers is rising rapidly. Therefore, it is important for European and U.S. companies to follow these developments, because they are certain to be affected in one way or another.

NTT

NTT still dominates Japan's telecommunication markets with about 73 percent of the long distance market and about 98 percent of the local-loop market. However, in the last few weeks, cable-TV operators have started to compete in the local-loop market as well. Japan's government decided on December 6, 1996 to divide NTT into four separate companies:

1. A long-distance operator, which will also offer international services (which NTT is at present not enabled to offer. As expected, NTT has entered the overseas (European, U.S., and Asian) markets (see sidebar, "NTT DoCoMo Set To Enter Foreign Markets").
2. Regional communications company: NTT West-Japan.
3. Regional communications company: NTT East-Japan.
4. A holding company, which will hold the shares of the previously mentioned daughters. For this, Japan's laws need to be changed, which at present do not enable this type of holding company.[2]

NOTE

The NTT group contains hundreds of specialized subsidiaries, which work in a wide variety of areas, including technology development, licensing, building management, and so on.

NTT DOCOMO SET TO ENTER FOREIGN MARKETS

NTT DoCoMo Inc. will launch its i-Mode mobile phone Internet service in Germany, the Netherlands, and Belgium in 2002 as part of a strategy to turn its wireless system into the mobile phone market's de facto standard. The company also plans to expand the service to other European countries, the United States, and Asia within the next year.

The overseas launch will be conducted with the cooperation of local partners, such as Dutch telecommunications company KPN Mobile N.V. and U.S. AT&T Wireless, which concluded business tie-ups with NTT DoCoMo in 2000. In Europe, NTT DoCoMo intends to concentrate on the three countries in which KPN already has a strong business base and expand later to the rest of the continent.

In Asia, NTT DoCoMo plans to initially offer the i-Mode service to Taiwan, Hong Kong, and South Korea, where mobile phones are in common use. In the United States, it will start service in part of the country by spring 2002 and expand later to other areas.

Before starting overseas operations, NTT DoCoMo will offer a new i-Mode-capable mobile phone to overseas users and establish a computer center in each country to provide the service. British mobile phone giant Vodafone, the largest rival of NTT DoCoMo, plans to expand its mobile phone network service to almost all of Europe by 2002. With the entrance of NTT DoCoMo in the European market, the competition between the two companies is likely to become fiercer.[3]

KDD

Until recently, KDD had the monopoly for Japan's overseas telecommunication and was, in exchange, forbidden to offer domestic services. This exclusion from the domestic market is about to end. KDD has already started to offer some domestic services. KDD laid 11,600 kilometers of deep-sea optical fiber cables around Japan. This project was called *Japan Information Highway* (JIH). It has 50 connections to shore stations, which are connected with the regional telecommunication networks associated with the regional electricity companies.

KDD, at present, is a public corporation owned by the government of Japan. On October 14, 1997, the *Ministry of Post and Telecommunications* (MPT) decided to privatize KDD during fiscal year 1998. KDD is open to foreign investment and does not need MPT approval for mergers or other business transactions any longer. These measures need changes of relevant laws by the Japanese Diet (Parliament). Legislation is being prepared to be submitted to the Diet.

New Common Carriers (NCCs)

The monopolies of NTT (domestic services) and KDD (international services) were lifted, and competition started in Japan's telecom markets. Today, more than 450 *new common carriers* (NCCs) exist, which compete with KDD and NTT with owned telecommunication infrastructure installations.

The strongest players are the three long-distance companies (DDI, Japan Telecom, and Teleway Japan). In addition, 13 regional telecommunications companies are associated with the regional electricity companies. In addition, there are two international NCCs: IDC and ITJ. ITJ and Japan Telecom (which is related to the railway group JR) merged. The new company's name is Japan Telecom Co., Ltd. DDI has an alliance with KDD. In addition, over 7,000 Type-2 carriers do not own their own circuits and usually offer various types of value-added services.

Communication Satellites

Two communications satellite companies are in Japan: JSAT and SCC. Both offer communication services, satellite TV, and digital services, including the MPEG2 format.

Cable TV as NCCs

Cable TV is relatively smaller than in the United States (10 percent in Japan versus 68 percent in the United States). Several cable TV operators have obtained the permission to offer telecommunication services in competition with NTT. Some have now started to compete with NTT by offering local-loop services, where NTT still has a de-facto monopoly.

WIRELESS AND CELLULAR PHONES

Mobile communications have recently exploded in Japan. Demand has been artificially held back by the regulators before. The number of mobile telephone users is now between 70 and 80 million (about 42 to 49 percent of Japan's population of 165 million, versus only about 10 to 14 percent of Germany's 120 million). Japan's mobile market is dominated by NTT DoCoMo with a market share of approximately 60 percent to 70 percent. The rest of the market is shared by several groups of mobile-NCCs.

Concerning the future, the Japanese government has decided to introduce the *Code Division Multiple Access* (CDMA) system as a standard for future cellular telecommunications. Also, a study group was formed to analyze the *Future Public Land Mobile Telephone Systems* (FPLMTS). NTT DoCoMo is very advanced in developing the so-called *wideband CDMA* next generation multimedia mobile systems in cooperation with a selected group of equipment manufacturers.

INTERNET

According to a recent survey by Nikkei Market Access, there were 13.0 million Internet users in Japan in 2000, or about 7.8 percent of Japan's population. Ninety-four percent were male, and only 6.0 percent were women! The number of users rose by 100 percent from 1997 to 2001.

At present, about 6,000 registered *Internet Service Providers* (ISPs) and about 1,200,000 Internet host computers exist. It is expected that Japan's Internet users will soon number about 60 to 70 million.

NTT and several NCCs have started to introduce new Internet services that are very aggressively priced and may provide tough competition for some of the ISPs, which have to pay for NTT connection services at present, because NTT has practically a monopoly in the local loop. The Japanese government has enabled Internet telephony, and several companies are testing these services.

TRENDS OF JAPAN'S CELLULAR MARKETS

Japan's cellular markets and investments total about $590 Billion and are growing at about 24-percent per year. Very soon, Japan's cellular markets will be almost completely deregulated and open, and all previous monopolies by NTT and KDD have been ended. Therefore, there will be very interesting opportunities for U.S., European, Canadian, and other companies, and many more will be affected in one way or another. Therefore, U.S., Canadian, and European companies are well advised to keep informed about the rapidly changing Japanese cellular market (see sidebar, "Cellular Markets in Japan").

As mobile-phone use explodes and phones get smarter, a race is on. Tech giants and small companies alike are competing to get their software accepted fastest and first. At stake, a slice of a pie potentially comparable to today's PC software market.

MOBILE PORTAL MARKETS

Mobile portals, which aggregate mobile content and services for mobile users, are forecast to be lucrative markets for those who can survive the competitive landscape. In Europe alone, the mobile portal market is likely to be worth between $6 billion and $11 billion a year in 2005. In the United States, the Strategis Group suggests that by 2006, there will be over 24 million mobile portal subscribers from the current 2 million in 2001 (see Figure 16-1).[5]

CELLULAR MARKETS IN JAPAN

Japan, the third largest cellular market in the world, is poised to be the global trendsetter in mobile data, 3G systems, and applications. The country is showing signs of rare opportunities for foreign players in its exploding cellular market—one that Pyramid forecasts will exceed 80 million subscribers by year 2005.

Cellular markets in Japan provide the insight you need to take advantage of the opportunities in this unique market. This incisive report includes detailed analysis of the telecommunications regulatory environment—its structure, processes, and outlook for the next five years. It provides comprehensive forecasts of access technologies paving the way for Japan's broadband future, analysis of the phenomenal success of i-Mode Internet-enabled mobile phones, an in-depth look at recent mergers and acquisitions, and and the Pyramid perspective on NTT following its July 1999 restructuring as a giant holding company. Other market players profiled include firms, such as KDD, DDI, Japan Telecom, Cable & Wireless IDC, Astel, MCI Worldcom, Level3, Lucent Technologies, Alcatel, Nortel, NEC, Fujitsu, Matsushita, and others. For those seeking guidance on developing competitive strategies in Japan, this chapter is not to be missed.[4]

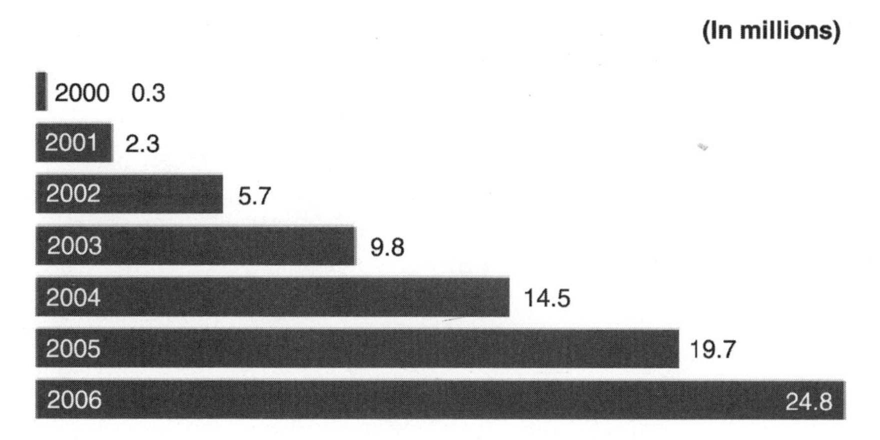

FIGURE 16-1 U.S. mobile portal subscribers: 2000 to 2006.

Those vying for a share of the mobile portal market will be coming to it from different angles. Mobile operators, such as DoCoMo in Japan and Vodafone in the United Kingdom, have already established large user bases for their mobile portals to support their voice and data services. Existing ISPs and Internet portals, such as AOL, MSN, and Yahoo!, which have established vast networks of content and services, are also offering their content to mobiles. Other players in the market include specialist content providers, such as news, banking, and retail companies, pure-play mobile portals that specialize in aggregating mobile content, and device manufacturers, such as Palm and Nokia, as shown in Table 16-1. [6]

The companies with the most to lose from the competitive mobile portal environment are the mobile operators. If mobile operators, such as Verizon, Cingular, and AT&T Wireless, are unable to offset the expensive costs of rolling out next-generation data networks with increased revenues from their subscribers paying for mobile content and services, they will find themselves struggling to balance their books. They must evolve from being just access providers of mobile telephony to service providers offering mobile content and applications to both consumers and businesses.

Those in the best position to take advantage of the competitive mobile portal environment are the established portals and specialist content providers. These companies, with a minimum of investment, will be able to distribute their existing content

TABLE 16-1 Major Mobile Portal Players Worldwide, 2001

MOBILE OPERATORS	EXISTING ISPs/ PORTALS	SPECIALIST CONTENT PROVIDERS	PURE MOBILE PORTALS	DEVICE MANUFACTURERS
NTT DoCoMo (i-Mode)	MSN Mobile	E*Trade	Halebop	Nokia
Vodafone (Vizzavi)	Yahoo!	Amazon	InfoSpace	Palm
Verizon	AOL	eBay	Eoom33	Ericsson

assets to mobile platforms and quickly take advantage of the enhanced data capabilities of next generation networks as they become available.

CONCLUSION

This chapter presented an overview of i-Mode markets. In conclusion, software companies see a wireless market potentially as valuable as that for personal-computer software.

Right now is a turning point for the mobile-telecoms world. This is a transition, and transitions are the best time to establish yourself for the next wave.

It's a second surge. In the past two years, the Internet technology called *wireless application protocol* (WAP) failed to deliver a decent Internet service to mobile handsets on global system for mobile or GSM standard, phone networks. GSM is used in most of the world, excluding Japan, South Korea, and parts of the United States. However, GSM networks are being upgraded again, even before much-vaunted 3G systems are installed. The upgraded GSM networks will mostly use the *General Packet Radio System* (GPRS), which offers a similar capability to i-Mode and other Japanese networks, enabling *always-on* connections to the Internet and faster access times—both weak links for WAP.

In short, the rest of the world (indeed the great bulk of the world's mobile-phone market) is about to catch up with Japan. Small outfits like Access are trying to push their software into the global big leagues, while technology heavyweights, such as Nokia, Matsushita, and Microsoft, are competing to sell their products as the world's mobile-software standard. Telephones are no longer just about talking; they're about data and the Internet too.

Wary of a Dominant Force

Call it the great software hand grab of 2001. About 360 million PC users are in the world. However, that's less than half the number of mobile-phone subscribers, and the phone market is expected to grow easily to over 2 billion within the next five years, according to a U.S. research company, the Yankee Group. Another U.S. research company, IDC, predicts new technology, such as improved handsets and GPRS, even without 3G, will still drive mobile operators' nonvoice revenues from $4.4 billion in 2001 to $55.9 billion by 2005.

That ubiquity offers a potentially powerful new market for developing technology and services, but cell phone-network operators and manufacturers are wary of vesting that power too closely in one place. The operators want the software on the phones to allow customers to access services quickly (be they e-mail, Web browsing, or voice calls) and for additional services to be developed quickly.

The devices (both the software and the hardware) are just the tools through which telecoms deliver. What's important is to have software on the phones that is well supported by developers so that they can innovate and build better products.

Ground zero in the emerging battle are those companies developing the *operating systems* (OSs) on the new devices—the central piece of software that enables other programs like e-mail, Internet browsing, address lists, and games to run. For now, the market is split into three broad segments: regular cell phones with simple Internet and e-mail connections, *smart phones* with sophisticated e-mail and browser functions, games-playing ability, and wireless hand-held organizers, such as the Palm or Handspring devices, which are close in ability to a notebook computer, but with added phone capability.

In the near term, much attention will focus on the organizer and smart phone segments, both of which will need sturdy and complex software-operating systems. Palm has already sold its OS on a Qualcomm cell phone, whereas Handspring (a Palm OS licensee) has added a GSM module onto its Visor organizer.

Microsoft, meanwhile, has already tailored its Windows OS into two forms—Pocket PC, which runs on hand-held organizers, and the Stinger OS for smart phones. Both OSs tie in tailored versions of Microsoft's Windows PC OS with its e-mail and Internet browser, and Microsoft's Internet server products that form part of the company's *.Net* strategy to rent its software services via the Internet.

Pocket PC is already catching up with market-leading Palm OS organizers, particularly with the eye-catching Compaq iPaq. Pocket PCs with integrated phones will soon hit the market from Mitsubishi subsidiary Trium and also from French company Sagem. Microsoft's Stinger will be available on the Z100 phone, made by British company Sendo, later this year and after that, from Samsung of South Korea, High Tech Computer of Taiwan, and Trium. Microsoft announced recently that it had taken a 10-percent stake in Sendo.

The mobile-phone industry is uneasy with Microsoft's move into its territory. Hence, none of the global leaders in the handset industry have embraced Microsoft's OS. Microsoft says the industry is at a turning point and that handset builders, until now, have held too much sway over operators. The company believes that the handset makers it has enlisted will take the market share away from the incumbents.

Indeed, wary of ceding domination of the mobile-phone industry to Microsoft, leading manufacturers Ericsson, Motorola, and Nokia acted to avert that in 1998: together with British hand-held computer-maker Psion, and later Japan's Matsushita Communications Industrial, which makes Panasonic phones, they formed Symbian, a British OS provider for mobile phones.

Symbian has developed its own OS, which it hopes will power Internet-enabled phones around the world. Symbian's new OS was released in June on a new Nokia gadget called the Communicator 9210. The Communicator is a large and expensive hybrid organizer, combining Internet access and voice phone. Symbian gains a royalty from every phone sold with its OS, so it is keen to move into the mass market. Although Symbian is in the top-end market right now, the vast majority of phones in use are cheaper and smaller, so they will be hoping to move into that market.

Symbian's presence gives many manufacturers ease of mind that their business won't be dictated by Microsoft. There isn't any situation in which Microsoft can be too dominant, because they have so many good competitors. Matsushita supports the

Symbian effort, but, like most manufacturers, the company will consider as many software options as possible.

Arakawa, meanwhile, is hoping Symbian and Microsoft don't move down the scale too fast. The company's survival depends on flying under the radar screen of the industry heavyweights and, as such, the company is mostly interested in the low end of the market.

In fact, Tokyo-based Access's browser software will run on Microsoft's Pocket PC OS and soon on Symbian's OS. If Symbian or Microsoft start to dominate this end of the market, they could fold into their OS offerings many of the software solutions Tokyo-based Access builds and sells separately, much as Microsoft did with its Internet Explorer browser for the PC. Access is wary of companies that might try to extend the OS to include browsers.

While handset OSs are the central battleground, important bridgeheads are also being formed by companies in the wireless-data server market—the backroom software that network operators need to create the services that users will access via their handsets. Key suppliers in this sector are Tokyo-based Access in Japan and Openwave of the United States, and just starting to roll out products is Microsoft.

Currently, Openwave claims it has 70 percent of the world market for providing software that links cell phones to the Internet, while Tokyo-based Access is the market leader in Japan. Because the market is shifting into a new 2.5G phase, past performance may not count for much—either way IDC predicts this wireless-infrastructure software market will be worth around $6 billion in 2004 from $217 million in 1999. Symbian won't address this part of the industry: they're still prohibited from selling infrastructure because their owners don't want to see lock-in client-server applications coming from one vendor.

Tokyo-based Access and U.S.-based Openwave have built their businesses largely on linking their Web browsers to their own server products. Openwave has until recently focused on WAP, whereas Access dominated i-Mode, but both have had to make their servers and browsers as universal as possible to appease operators.

Both companies believe their telecoms experience will help them fend off Microsoft's push into this market. Phone users demand much faster response and reliability than PC users.

Operators are most interested in whether the software is a good base for developing new services. Along with the GPRS upgrades to the GSM networks, other new technologies, in particular small downloadable programs written in Java, are quickly becoming very popular.

Microsoft, Symbian, and Palm all emphasize their existing base of third-party developers and software partners. Regardless of who emerges as a dominant player, the exciting opportunity for most software companies is the potential of a new market. I-Mode's popularity in Japan has software houses salivating as GPRS and 3G markets come online.

Finally, the software players' position is rising. Customers want more functions. The industry is concentrating on making better handsets and the components that make them attractive.

END NOTES

[1]John R. Vacca, *The Cabling Handbook* (2nd Edition), Prentice Hall, Upper Saddle River, N.J., 2001.

[2]*Japan's Internet M-Commere Markets*, Eurotechnology Japan K.K., Parkwest Building 11th Floor, 6-12-1 Nishi-Shinjuku, Shinjuku-ku, Tokyo 160-0023, Japan, 2001.

[3] Synchrologic, 200 North Point Center East, Suite 600, Alpharetta, GA 30022, 2001.

[4]The Economist Intelligence Unit Limited, The Economist Building, 111 West 57th Street, NY, NY 10019, USA, 2001.

[5]Ben Macklin, *Mobile Portals*, EMarketer, 821 Broadway, New York, NY 10003, 2001.

[6]Ibid.

CHAPTER SEVENTEEN

OVERVIEW OF THE JAPANESE CELLULAR MARKET

In recent years, Japan has become a dynamic, commercial magnet. Companies from many countries and industries have entered Japan's growing markets, establishing representative offices and building factories. This trend can especially be seen in the Japanese cellular industry. American and European telecommunications giants are vying for contracts to develop Asia's inadequate telecommunications infrastructures, with hopes to construct networks and sell equipment.

Nippon Telegraph and Telephone (NTT DoCoMo) is a Japanese telecommunications company that was established as a public company in 1952, after operating as a governmental organization since the 1890s. It was privatized in 1985 and offers virtually every aspect of cellular services, although it does not manufacture equipment nor handle international calls. NTT DoCoMo, like its Western counterparts, has also been active in entering the larger Asian cellular market to construct telecommunications networks and related systems. By focusing on its regional strength and Asian ties, NTT DoCoMo has been able to differentiate itself from its Western competitors. With this focus, NTT DoCoMo is now working hard to expand its presence in the Asian cellular marketplace.

RECENT CHANGES IN THE JAPANESE CELLULAR MARKETPLACE

In the last 10 years, dramatic changes have taken place in the Japanese cellular market. Four key elements for the changes have taken place:

- Customers are becoming global. Businesses have been expanding their operations to other countries, creating truly multinational businesses. They require telecommunications companies that provide cross-border services. Some developing countries, especially in Asia, have emerged as potentially big telecommunications customers. When it comes to individual users, they want more personalized services.[1]
- Communications technology has advanced. Remarkable and rapid progress of such technologies as digital, wireless, optical fibers, and the Internet has led to a convergence of services and businesses in telecommunications.
- Service providers have had to keep up with the changes in customers and technology. Telecommunications giants successively formed global alliances and joint ventures to find new market opportunities and try unfamiliar technologies and services.
- These changes affect government officials' public policy decisions in the area of telecommunications. To improve social welfare, policymakers and regulatory authorities in most of the industrialized countries, and in some emerging cellular markets, have attempted to stimulate their cellular markets by privatizing telecommunications operators and introducing competition.[2]

Prior to the 1980s, telecommunications services were provided by governments, public entities, or highly regulated monopolies. Economists generally believed that telecommunications services were a natural monopoly with a tremendous capital investment as an entry barrier. Customers had to be content with a *plain old telephone service*.

The remarkable progress in digitalization, high-capacity transmission, *Very Large Scale Integration* (VLSI), and wireless technology has so significantly changed the economics of telecommunications that it no longer has to be a government-owned enterprise, and few would now consider it to be a natural monopoly. Competing telecom networks can now be built with a significant, but not prohibitively expensive, capital investment (see sidebar, "Dealing with NTT DoCoMo"). There is also a growing customer sophistication and increasing demands for more advanced communications services and even multimedia.

DEALING WITH NTT DOCOMO

Leading mobile phone operator NTT DoCoMo, may be targeted by Japan's *Fair Trade Commission* (FTC) policy guideline to promote competition in Japan's cloistered cellular market. FTC referred to the company by indicating that moves by a mobile phone operator to defer offering the technology it jointly developed with telephone set makers could be problematic in light of the Antimonopoly Law.

The five-point guideline jointly drafted by the FTC and the telecom ministry is aimed at clarifying competition rules and complementing the Antimonopoly Law in governing telecom carriers. The guideline, which will be formalized in 2002 at the earliest, prohibits dominant carriers from overshadowing newcomers that use their facilities and forces carriers possessing utility poles, channels, and other base facilities to make them accessible to other companies.

It also prohibits carriers from interfering in business tie-ups by related program content providers with other companies. The draft, however, does not address the high-profile issue of setting a limit on the stockholdings by holding company Nippon Telegraph and Telephone Corporation in NTT DoCoMo and other dominant affiliates. The guideline marks the first policy coordination effort between the FTC and the Public Management, Home Affairs, Posts and Telecommunications Ministry, since the fair trade watchdog came under the ministry's jurisdiction in the January 6, 2001 realignment of government bodies.[3]

Major Japanese Cellular Carrier

Most of the primary Japanese cellular carriers are big businesses, due to the large capital investment required to serve the cellular marketplace as a whole. AT&T teamed with *Kokusai Denshin Denwa* (KDD), a Japanese international cellular company, Singapore Telecom, and invited numerous European and Asian telecommunications carriers to their group. This group is called *World Partners*.

As domestic demand in the cellular markets of developed countries becomes saturated, carriers have sought to enter foreign markets. Prime targets are developing countries, where the telecommunications infrastructure is incomplete, and large projects are being considered. One of the most potentially lucrative areas is Asia, where booming economies can sustain growth and fund the necessary large capital investments.

Current Cellular Market Conditions in Japan

Taking advantage of this high economic growth, Japan is eager to attract foreign investment to quickly complete the construction of communications infrastructures and to catch up with more industrialized countries. Many Japanese cellular companies hope that competition and foreign investment will stimulate their markets.

It should be mentioned that the *National Information Infrastructure* (NII) and the *Global Information Infrastructure* (GII) concepts, popularized by the U.S. government, helped Japan understand the importance of telecommunications development. It is now well understood that telecommunications is a necessity to secure economic development and increase public welfare.

The Japanese market is very promising. It is estimated that the 1998 to 2002 annual growth rates will average 66 percent for cellular phones and 70 percent for wireline telephones. This compares with U.S. figures of 24 percent in wireless[4] and 2 percent in wireline, and European figures of 20 percent in wireless and 1 percent in wireline.

JAPANESE CELLULAR CARRIERS AND THEIR ACTIVITIES IN THE ASIAN MARKETPLACE

In Japan, two public corporations, NTT DoCoMo and KDD have provided telecommunications services since 1952. NTT DoCoMo handles domestic telecommunications services, and KDD handles international services in the Japanese market. After the privatization of NTT DoCoMo and the opening of the Japanese cellular market in 1996, fierce competition has occurred, primarily in the long-distance and international cellular markets.

As a result of continuous efforts by its carriers, the Japanese market enjoys relatively high-quality services. The network digitalization rate is 83 percent in Japan, in contrast to 76 percent in the United States and 66 percent in the United Kingdom. Also, the number of phone line connection faults per 100 lines per year is 3 in Japan, 9 in the United States, and 28 in the United Kingdom. Phone lines per employee, which indicates employee productivity, is 359 in Japan, 336 in the United States, and 264 in the United Kingdom.

Japan's infrastructure is highly developed. A wide variety of services are available in Japan, from basic switched voice services to advanced high-speed data services. Cellular services are available throughout Japan, and last year, PHS (Japan's version of PCS) was introduced in several metropolitan areas.

When it comes to foreign investment, however, Japanese telecommunications carriers are behind. The cumulated overseas investment by NTT DoCoMo, from 1995 to 2000, amounted only to $12 billion, while British Telecom and AT&T invested $215 billion and $103 billion, respectively. BellSouth, one of the telecos, made $25 billion worth of foreign investment during the same period. This comparison indicates how inactive NTT DoCoMo, as one of the Japanese cellular carriers, has been in foreign business.

Resistance to foreign investment exists for two reasons. One reason is that the Japanese cellular companies and the government have found it unnecessary to enter into foreign markets.

After World War II, the Japanese telecommunications system was destroyed. The Japanese government, together with its telecommunications operators, were absorbed in restoring the damaged system for more than 45 years. During this period, they lost their focus on what was happening in the worlds cellular market.

Another reason is that the government had regarded telecommunications as highly important because it significantly relates to the public interest and is vulnerable in that it also relates to national security. Therefore, the government has established a web of regulations and made minute administrative guidance to keep its cellular operators away from any possible trouble. Foreign investment restrictions and the distinction of business areas between NTT DoCoMo and KDD exemplify these rules.

For the Japanese telecommunications industry to keep up with global trends, the first thing it needed to do was to bring more thorough deregulation and competition to its domestic market. The global business users require cross-border support, and Japan's Asian allies expect Japan to support them economically and technologically.

Japanese industry benefits from being an integral part of the growing Asian market, and it shares a common responsibility in seeking to attain Asian prosperity and unity. Part of Japan's role in supporting such a trend is to respect other countries' ideas and help them achieve their goals as efficiently as possible.

Another international responsibility that Japan must accept is that in order to be a player in world markets, Japan must play by world rules. World trading rules are governed by the notion of *national treatment* and reciprocity. NTT DoCoMo advocates a Japanese market open to competition and foreign investment.

NTT DoCoMo's Current Overseas Business Activities

Though NTT DoCoMo is primarily confined to domestic business activity in Japan, it is also allowed to enter overseas business so as to contribute to the construction of worldwide advanced telecommunications infrastructures. Particularly in

Asia, NTT DoCoMo has supported various countries by dispatching its experts for assistance through Japan's international cooperation program.

Recently, NTT DoCoMo has been active in cooperation with Asian countries, such as Thailand, the Philippines, and Indonesia, helping to establish their telecommunications networks. NTT DoCoMo has established representative offices in Shanghai and Hong Kong and is actively involved in the bidding for basic telephone service in India. In Vietnam, one of NTT DoCoMo's subsidiaries, NTT DoCoMo International is to participate in basic telephone network construction in Hanoi. Other examples include the followiing:

- The first major project was in Thailand. In November 1993, NTT DoCoMo joined *Thai Telephone and Telecommunications* (TT&T) as a strategic partner and became a 20 percent equity holder. TT&T's goal has been installing one million telephone lines throughout Thailand except the Bangkok metropolitan area. Thanks to continuous effort and excellent collaboration with the *Telephone Organization of Thailand* (TOT), in December 1996 the number of line connections reached over 500,000, and the number of installations is now over 1,940,000 lines.
- The second major project has been in the Philippines. In January 1994, Smart Communications Inc., was established by investors of Hong Kong and the Philippines to provide mobile telephone and international telephone service in the north and central part of Luzon island and the south part of the Manila metropolitan area. NTT DoCoMo acquired a 15 percent share of Smart in 1995 and has given technical and operational support since then. In November 1996, Smart successfully launched basic telephone service.
- The third project is the joint venture in Indonesia to install 500,000 telephone lines in central Java. The joint venture was established by NTT DoCoMo, PT Telecom of

Indonesia, Telestra of Australia, Indosat of India, and various local investors. NTT DoCoMo holds 15 percent of the stock and is providing full technical and operational support.[5]

SOME INSIGHTS ON JAPANESE CELLULAR BUSINESS IN ASIA

The following are some important aspects in the Japanese cellular business with Asian countries.

- Concerned parties have to share their views even if their opinions differ. They should not lose sight of the common goal. Candid talk is essential.
- To keep the project on schedule, quick and timely decision making by top management is necessary.
- To manage constantly changing business conditions, it is sometimes necessary to improve or modify the rules. Flexible operations can overcome problems.
- Strong support from the government is important in that we may sometimes need to ask the authorities to ease regulations or accelerate approval periods for services.
- The operator's role should not be overlooked. The operator has to provide a general feeling of assurance, though not a guarantee, to the project. Professional analysis and constant support for the overall project is indispensable for operators.[6]

CONCLUSION

This chapter presented an overview of the Japanese cellular market. In conclusion, Asia is filled with opportunity. However, the local players will become either your intimate business partners and your key to valuable markets, or they will become your fiercest competitors. The economy of each country in the region is growing at a remarkable pace. Whether you will be successful in business in Asia depends upon how carefully and quickly you grasp the present condition of each country's business and key players there. One should watch for a crack in the market and seek a niche, because the established players are going to be there for a while.

END NOTES

[1]Keisuke Nakasaki, "Challange In Asian Telecommunications Market: A Japanese Perspective," NTT DOCOMO America, Inc., 101 Park Avenue, 41st Floor, New York, New York 10178, 2001.

[2]Ibid.

[3]*FTC Suggests Dealing with NTT DoCoMo*, AnywhereYouGo.com, 3000 Waterview Parkway, B2E14, Richardson, TX 75080, 2001.

[4]John R. Vacca, *Wireless Broadband Networks Handbook: 3G, LMDS, and Wireless Internet*, McGraw-Hill, New York, 2001.

[5]NTT DOCOMO America, Inc., 101 Park Avenue, 41st Floor, New York, New York 10178, 2001.

[6]Ibid.

CHAPTER EIGHTEEN

THE JAPANESE MOBILE DATA MARKET

Before i-Mode service was launched by NTT DoCoMo, a cellular phone was just a connection tool for Internet users to access the Internet with their portable PC, but no more. Now Japanese mobile data users access the Internet with the tiny browser phone using only their thumb. No PC is necessary.

The big success of i-Mode service is winning attention not only in Japan, but also abroad. It is because i-Mode is the first successful IT-world service originated in Japan. Before i-Mode, Japanese consumers browsed the Internet using desktop systems powered by Windows 95, 98, 2000, or XP, with Internet Explorer or Netscape—all American innovations. I-Mode is a purely Japanese product.

KEY FINDINGS

NTT DoCoMo has been continuously upgrading i-Mode services. Full-color terminals are now available, music distribution is planned in the foreseeable future, and DoCoMo is scheduled to release Java-based terminals by the end of 2002.

Their ultimate goal is *third-generation* (3G) cellular services. DoCoMo is going to launch its 3G service, an advanced implementation of i-Mode, in 2002.

When DoCoMo launched i-Mode in February 1999, virtually nobody at DoCoMo expected such a big success. As of this writing, out of 70.6 million DoCoMo users, 33 million had signed up for i-Mode services. The carrier is now expecting an additional 37 million i-Mode subscribers by the end of the fiscal year 2002.

Following i-Mode's big success, other carriers started providing Internet access services. DDI Cellular Group is offering a service called *EZWeb*, IDO is providing *EZAccess*, and J-Phone Group is offering *J-Sky*. All told, Japanese mobile Internet data users exceed 30 million as of this writing.

Most of the Internet content is being provided free of charge, but much of i-Mode's (and other mobile Internet data services') content is provided for a small fee—typically ¥100 to ¥300 ($1 to $3) per month. Telecom carriers are collecting these fees with monthly connection charges instead of content providers.

Content providers can charge end-users in Japan, because Japanese carriers limit the number of content on their platforms. Each of the Japanese carriers selects a certain number of official content providers based on their own standards. Then, only these selected official sites are enabled to put their content on the carriers' portal sites. Carriers collect fees only from their official content sites.

I-Mode has 90 percent of the mobile Internet data services market. The strength of i-Mode is its abundant content. As of this writing, i-Mode has 723 official sites. In addition, more than 30,000 unofficial sites are online. Competition runs high for the limited *official* status, and content providers jockey for the DoCoMo seal of approval. Because DoCoMo closely examines all of them and selects only very attractive content, their services win more users.

As i-Mode grows, so does NTT DoCoMo's customer base. As of this writing, the number of subscribers reached 52.6 million. However, the total increase of cellular users has been slowing down, as the user number is reaching the saturation point. It is said the number of cellular users will reach a ceiling of 100 million (67 percent of the total Japanese population) by 2010. In order to break the limit, NTT DoCoMo is trying to include a cellular phone (or an equivalent device) in anything that moves (and some things that don't), including pets, vending machines, and cars (see sidebar, "DoCoMo in KPN").

NTT DoCoMo indicated that they could sell 582 million cellular phones (or equivalent devices) in Japan by employing the strategy. These new *users* will expand mobile data traffic rather than voice. The mobile data communications is expected to occupy 80 to 90 percent of the total traffic by 2010. However, revenues from mobile data services are not as profitable as voice. So, they may have to seek out additional revenue sources, such as advertising in the future.

DoCoMo is scheduled to launch its 3G services based on *Wideband Code Division Multiple Access* (W-CDMA) at the end of 2002. The big success of i-Mode clearly demonstrates the big business opportunity of 3G services. The 3G service is expected to have a rosy, futuristic life. Because the system is internationally standardized, users will be able to use their phone anywhere in the world. Because the 3G will provide 386-Kbps to 2-Mbps mobile data transmission services, users will be able to enjoy full multimedia services like motion pictures or a TV phone.

With *second-generation* (2G) mobile data transmission, Japan employed a unique standard called *PDC*, which isolated Japanese carriers and venders from world cellular market vendors. Learning from its experience with 2G, NTT DoCoMo focused all of its effort to make its original 3G technology, called W-CDMA, compliant with the international standard for 3G. DoCoMo is now ready to take the leadership in 3G, riding on the big success of i-Mode.

DOCOMO IN KPN

NTT DoCoMo, Japan's largest mobile-phone operator, agreed recently to pay around $5 billion ($4.5 billion) for a 15 percent stake in the cellular arm of Dutch telecom market leader Royal KPN, paving the way for further expansion in Europe. DoCoMo has made no secret of its global ambitions, and analysts indicate the deal will provide a bridgehead to bring NTT's new-generation cellular technology into the European market and could form the basis for a joint bid with KPN for Orange, Britain's third-ranked cellular player.

KPN and DoCoMo, the world's second-largest cellular operator after Vodafone AirTouch, have previously been linked to separate bids for Orange, which is being sold by *Vodafone* (VOD). The cutting-edge technology that DoCoMo will bring to the partnership will assist KPN Mobile in staying ahead in the introduction of innovative next-generation mobile services.

DoCoMo already has 33 million users of its i-Mode Internet-enabled phones in Japan and, together with KPN, plans to develop 3G services, which provide fast Web access. The agreement with DoCoMo follows the collapse of KPN's merger talks with Spain's Telefónica, amid disagreements over the Dutch government's involvement in a merged venture. The breakdown prompted speculation that KPN would seek other partners to help fund its expansion of new-generation wireless services[1] and a possible bid for Orange.

KPN indicated it plans an initial public offering of KPN mobile shares when the DoCoMo deal is completed, subject to regulatory approval. The terms of DoCoMo's purchase of a 15-percent stake imply the whole of KPN Mobile is worth about $33 billion.

EXPANSION CONCERNS DOG NTT

The deal is DoCoMo's largest investment outside Japan. The new partners have indicated they would create a joint task force to accelerate the introduction of data and Internet services in Europe. Analysts have indicated the KPN investment was a boost for the Japanese company's plan to promote its CDMA technology as the global industry standard for the next generation of mobile phones.

KPN Mobile, the largest player in the Dutch cellular market, has expanded by buying a stake in Germany's E-Plus and already operates a joint venture with Orange in Belgium. KPN is also bidding for a 3G mobile license in Germany.

NTT underlined its global ambitions recently when NTT Communications, the group's third arm, agreed to pay $5.5 billion for U.S. Web site designer Verio and roll out a network of data centers to provide Web services for large companies. Credit-rating agency Standard & Poor's placed DoCoMo and NTT on its *Creditwatch* list with a view to a possible downgrade after DoCoMo in 2000 paid $521 million for a 19-percent stake in Hutchison Telephone, Hong Kong's largest cellular operator, which had previously sold its stake in Orange to Mannesmann.

VOD is selling Orange to meet demands from European regulators after its takeover of Germany's Mannesmann gave it a second U.K. mobile-phone network. *France Telecom* (PFTE) has been seen as the front-runner for Orange; its shares were down 4 percent.

The success of i-Mode revealed certain limitations of the mobile Internet data for end users. Entering mobile data can be difficult—only well-trained youths can input mobile data using only their thumb. The Japanese use a variety of characters such as kana, katakana, Chinese characters, and the alphabet, which further complicates matters. Mobile data transfer speed is limited to 9.6 Kbps, and the screen of most devices is rather tiny (the largest screen for an i-Mode terminal can demonstrate merely 10 letters × 10 lines).

Many experts said that i-Mode could only have succeeded in the Japanese market because the Japanese like a tiny tool, are sensitive to new trends, and neither the Internet nor the PC has been used widely in Japan. The i-Mode business model will not succeed in the United States, where both the Internet and PCs are widely used already. Americans do not like to view the Internet on such a tiny screen and they do not like to input mobile data with just their thumb either (see sidebar, "America's Way Behind Others when It Comes to Web Phones").

AMERICA'S WAY BEHIND OTHERS WHEN IT COMES TO WEB PHONES

Big is always better. So goes the American creed we have long subscribed to. That's why most of us haven't deigned to try out that hybrid grotesquerie: the Internet phone. Web surfing on one of those dinky little screens seemed about as much fun as body surfing in the bathtub.

Nevertheless, we were all wrong. Unlikely tool of revolution though it appears to be, the Internet-connected gizmo is going to usher in truly ubiquitous computing. We have now seen the future, or rather 33 million Japanese have seen the future, and, amazingly, it works. Alas, the Japanese have a more advanced network than Americans do. The American version of a Web phone (its infrastructure) is creaky and slow. America's wireless carriers still use circuit-switched networks, which owe more in the design to the manual telephone switchboards of yore, than to today's Internet architecture. Circuit-switching makes a Web phone's connection to the Net expensive and cumbersome.

Japan's NTT DoCoMo, however, has a snazzy system, the first packet-switched data network, which uses the same principle that the Internet uses. In Japan, a Web phone appears to the user to be *always on*, able to dispatch or receive data instantly. Rather than attempting to make an Internet phone do what it simply cannot do (serve as a replacement for the PC for navigating the Web), DoCoMo's i-Mode service offers subscribers other ways to use the Internet's connectivity. Instant messaging, for example, has become the rage among Japanese teenagers: who'd a think that a cell phone, a very serviceable tool for chitchat, would instead be used for text chat?

I-Mode succeeded in Japan, where carriers have a bigger influence than venders. In Japan, carriers purchase terminals from venders and sell these terminals at a low price (occasionally free), paying high incentives to retailers. Carriers recover the cost with monthly communications fees.

Due to the system, young people, such as high school kids, have embraced i-Mode. However, in European countries or in the United States, venders have more influence than carriers, and they sell their terminals at a retail store. So, the Japanese business model does not work in these countries.

NOT ALL TALK

Services that offer a daily cartoon, weather reports, horoscopes, train schedules, bank account information, and stock quotes are also popular. In just 36 months, i-Mode has attracted 33 million paying subscribers—a phenomenon that American Web content providers must view with doleful envy. There's more than teen talk here. Japan Airlines already sells 40,000 tickets a month on the service, a feat enabled by designers who figured out ways to let users get to schedules in two clicks. By contrast, an American punching a Web phone needs seven just to get a flight number. AOL has taken notice and recently reached an agreement to set up a joint venture in the United States.

The Japanese have gotten this off the ground without even having a truly fast wireless data network in place. That will come with the 3G network, which DoCoMo will introduce in 2002. Europe will soon deploy fast wireless too. Rights to build 3G networks drew eye-popping bids in European spectrum auctions recently, most notably $35.5 billion in the United Kingdom and $42 billion in Germany.

Trailing way, way behind the rest of the industrialized world in this business is the United States, which will be lucky if they get started in 2006. The reasons (all political) would be amusing were it not for a nagging suspicion that 3G is going to be the crucial piece of infrastructure for not just cell phones, but PDAs, laptops, and automobiles.

American carriers are eager to build out 3G, and Congress is eager to get a hold of the billions that will be paid for the licenses. However, the *Federal Communications Commission* (FCC) has had to postpone the auctions twice because of the disproportionate power exercised by the television broadcasters lobby, a group that members of Congress always treat with solicitous care. Some 240 broadcasters happen to have rights to channels 60 to 69, and they have been given their moving notices by the government as part of the transition to digital television. Those channels also occupy the spectrum that carriers need for 3G. Seeing an opportunity to reap a windfall before they have to move anyhow, these station owners are demanding that they separately be paid billions to vacate early. No one in the government has been willing to invoke the Radio Act of 1927, which clearly provides for the reassignment of spectrum to *promote public convenience or interest*. Continued reruns of *Dr. Quinn: Medicine Woman*, instead of deployment in this decade of a 2 Mbps 3G wireless network? This is not a case study of policy setting in Washington at its finest hour.

continues

continued

The carriers balk at paying extortion, Congress and the FCC shrug as if helpless to intervene, and 3G remains on the shelf. Meanwhile, stations will continue beaming reruns of *Touched by an Angel.*[3]

CONCLUSION

This chapter presented an overview of the Japanese mobile data market. In conclusion, when i-Mode makes the transition to 3G, some problems may be solved, such as the slow mobile data transmission speed. Some new technologies, such as Bluetooth, may solve the input trouble; some by enabling a separate input device for the cellular phone.

Finally, the highest hurdle might be a psychological barrier. By using a concept that does not consider i-Mode as a telephone apparatus, i-Mode marketers have a breakthrough. When Americans acquire a more flexible concept, then browser phone services like i-Mode might be successfully implemented in the United States.

END NOTES

[1]John R. Vacca, *Wireless Broadband Networks Handbook: 3G, LMDS, and Wireless Internet*, McGraw-Hill, New York, 2001.

[2]Randall E. Stross, *DoMoCo In KPN Mobile Tie*, CNN America, Inc., An AOL Time Warner Company, One CNN Center, Atlanta, GA 30348, 2001.

[3]Ibid.

PART VI

ADVANCED I-MODE AND FUTURE DIRECTIONS

CHAPTER NINETEEN

THE FUTURE OF I-MODE

DoCoMo will introduce the *third generation* (3G) service with a starting speed of 64 Kbps in 2002 and boost the speed to 2 Mbps, which is 200 times faster than current CDMA, by 2003. In the 3G, users can watch TV, listen to music, and play games with a cell phone.

DoCoMo plans to overhaul i-Mode to fit *wideband CDMA* (W-CDMA). DoCoMo proposed to the *Wireless Application Protocol* (WAP) forum that a standard combining WAP and HTML be adopted for the next-generation system. For 3G, DoCoMo may go with whatever emerges as the main standard.

DoCoMo launched the *Joint Initiative for Mobile Multimedia* (JIMM) to discuss developing services for 3G services. Vodafone, British Telecom, France Telecom, AT&T Wireless, SK Telecom, and Singapore Telecom have joined this forum. The forum is supposed to adopt a system like i-Mode. I-Mode is heading to be *international Mode*.

However, problems do exist. Network problems recently have delayed the launch of 3G wireless services. Let's see why that is.

THE FUTURE: THIRD GENERATION

A small island in the Irish Sea and the island country of Japan are vying to become the first to host 3G wireless services, which bring multimedia applications to mobile phones with the i-Mode system. Japan's NTT DoCoMo was scheduled to introduce 3G in May 2001, in Tokyo, Yokohama, and Kawasaki, but postponed its full-scale commercial rollout to January 2002 to iron out problems with the handsets and with the network. If problems persist, the latest rollout date may fall by the wayside also.

DoCoMo is working out the bugs in a small pilot project in Tokyo that's limited to 5,600 customers. DoCoMo scaled back its May 30, 2001 launch after early users reported the handsets froze and had to be reset. Only 50 percent of attempts to connect to the network were successful and batteries lasted less than a day.

DoCoMo Draws Close

In Japan, NTT DoCoMo is dealing with the disappointment of delaying the commercial launch of its 3G service, marketed as *freedom of mobile multimedia access* (FOMA). Nevertheless, the company has selected *monitors* who are evaluating the service before the rollout. DoCoMo is providing free handsets and is waiving monthly fees for the monitors. It has received nearly 260,000 requests from prospective monitors, but is distributing only 5,600 3G handsets: 2,500 to individuals and 3,100 to corporate subscribers.

DoCoMo is offering three models of 3G phones: an upgraded model of its current mobile phone featuring higher-quality audio, a *visual* model with a video screen, and a *datacard* version for high-speed data transmission. Sixty percent of those applying for handsets want the visual model. However, that version's debut is running behind the others because of debugging delays.

DoCoMo recently announced that the network is becoming more stable and that DoCoMo plans to meet its January 2002 target date for a full-scale rollout. The two major problem areas with the network are the switching system that controls the connections, and the *handovers* that occur when a user moves from one base station area to another. So, as the race to be first in 3G enters its last lap, the Japanese carriers are being cautious.

MAKING THE QUANTUM LEAP TO 4G

Although 3G hasn't quite arrived, DoCoMo designers are already thinking about *fourth generation* (4G) technology. With it comes challenging RF and baseband design headaches.

Cellular service providers like DoCoMo are slowly beginning to deploy 3G cellular services (see sidebar, "Fourth Generation [4G] Broadband Mobile"). As access technology increases, voice, video, multimedia, and broadband dataservices are becoming integrated into the same network. The hope once envisioned for 3G as a true broadband service has all but dwindled away. It is apparent that 3G systems, whereas maintaining the possible 2-Mbps data rate in the standard, will realistically achieve 384-kbps rates. To achieve the goals of true broadband cellular service, the systems have to make the leap to a 4G network.

This is not merely a numbers game. 4G is intended to provide high speed, high capacity, low cost per bit, IP-based services.

The goal is to have data rates up to 20 Mbps, even when used in such scenarios as a vehicle travelling 200 kilometers per hour. New DoCoMo design techniques, however, are needed to make this happen, in terms of achieving 4G performance at a desired target of one-tenth the cost of 3G.

The move to 4G is complicated by attempts to standardize on a single 3G protocol. Without a single standard on which to build, DoCoMo designers face significant additional challenges. Table 19-1 compares some of the key parameters of 3G and 4G (4G does not have any solid specifications as of yet, so the parameters rely on general proposals).[1] It is clear that some standardization is in order.

FOURTH GENERATION (4G) BROADBAND MOBILE

At present, the download speed for i-Mode data is limited to 9.6 Kbps, which is about six times slower than an ISDN fixed line connection. However, in actual use the data rates are a lot slower, especially in crowded areas or when the network is *congested*. For 3G, new network hardware and software will be installed that will increase the data rates by a large factor (in the final stages up to 200 times faster). 3G in Japan will be introduced in stages starting early 2002. Japan will be the world's first country to introduce 3G. 4G mobile communications will have even higher data transmission rates than 3G. 4G mobile data transmission rates are planned to be up to 20 Mbps.

INTRODUCTION OF 4G (BROADBAND) MOBILE COMMUNICATIONS IN JAPAN

Initially, DoCoMo planned to introduce 4G services around 2010. Recently, DoCoMo announced plans to introduce 4G services from 2006 (four years earlier than previously planned).

THE DATA RATES FOR 4G (BROADBAND MOBILE) IN JAPAN

At present, (2G) impedance data rates in Japan are up to 9.6 Kbps, but are usually a lot slower. For 3G data, rates will be up to 2 Mbps (approximately 200 times faster). However, 2 Mbps will only be reached later in time. Initially, the data rates will be lower. For 4G data rates, up to 20 MBps per second are planned. This is about 2,000 times faster than present (year 2001) mobile data rates, and about 10 times faster than top transmission rates planned in the final build out of 3G broadband mobile. It is about 10 to 20 times faster than standard ASDL services, which are being introduced for Internet connections over traditional copper cables at this time (2001).[2]

NOTE:

Impedance is the combined effect of resistance, inductance, and capacitance on a signal at a particular frequency.

TYPE OF SERVICES 4G WILL ENABLE

Of course, it is impossible to predict technology developments and the evolution of culture and customer needs. 4G, in principle, will enable high-quality, smooth video transmission.[3]

TABLE 19-1 Key Parameters of 3G and 4G Systems

	3G	4G
Frequency band	1.8 to 2.5 GHz	2 to 8 GHz
Bandwidth	5 to 20 MHz	5 to 20 MHz
Data rate	Up to 2 Mbps (384 Kbps deployed)	Up to 20 Mbps
Access	W-CDMA	MC-CDMA or OFDM (TDMA)
Forward error correction	Convolutional rate $1/2$, $1/3$	Concatenated coding scheme
Switching	Circuit/packet	Packet
Mobile top speeds	200 km/h	200 km/h

MULTICARRIER MODULATION

To achieve a 4G standard, a new approach is needed to avoid the divisiveness you've seen in the 3G realm. One promising underlying technology to accomplish this is *multicarrier modulation* (MCM), a derivative of frequency-division multiplexing. MCM is not a new technology; forms of multicarrier systems are currently used in DSL modems and *digital audio/video broadcast* (DAB/DVB). MCM is a baseband process that uses parallel equal bandwidth subchannels to transmit information. Normally implemented with *fast fourier transform* (FFT) techniques, MCM's advantages include better performance in the *intersymbol interference* (ISI) environment and avoidance of single-frequency interferers. However, MCM increases the *peak-to-average ratio* (PAVR) of the signal, and to overcome ISI, a cyclic extension or guard band must be added to the data.

Any increase in PAVR requires an increase in the linearity of the system to reduce distortion. Proposed approaches to reduce PAVR have consequences, however. One such technique is clipping the signal; this results in more nonlinearity. Linearization techniques can be used, but they increase the cost of the system, and amplifier backoff may still be required.

Cyclic extension works as follows: If N is the original length of a block, and the channel's response is of length M, the cyclically extended symbol has a new length of $N + M - 1$. The image presented by this sequence, to the convolution with the channel, looks as if it was convolved with a periodic sequence consisting of a repetition of the original block of N. Therefore, the new symbol of length $N + M - 1$ sampling periods has no ISI. The cost is an increase in energy and uncoded bits added to the data. At the MCM receiver, only N samples are processed, and $M - 1$ samples are discarded, resulting in a loss in *signal-to-noise ratio* (SNR).

Two different types of MCM are likely candidates for 4G as listed in Table 19-1. These include *multicarrier code division multiple access* (MC-CDMA) and *orthogonal frequency division multiplexing* (OFDM), using *time division multiple access* (TDMA).

NOTE

MC-CDMA is actually OFDM with a CDMA overlay.

Similar to single-carrier CDMA systems, the users are multiplexed with orthogonal codes to distinguish users in MC-CDMA. However, in MC-CDMA, each user can be allocated several codes, where the data is spread in time or frequency. Either way, multiple users access the system simultaneously.

In OFDM with TDMA, the users are allocated time intervals to transmit and receive data. As with 3G systems, 4G systems have to deal with issues of multiple access interference and timing.

Differences between OFDM with TDMA and MC-CDMA can also be seen in the types of modulation used in each subcarrier. Typically, MC-CDMA uses *quadrature phase-shift keying* (QPSK), although OFDM with TDMA could use more *high-level modulations* (HLMs), such as *multilevel quadrature amplitude modulation* (M-QAM) (where $M = 4$ to 256). However, to optimize overall system performance, adaptive modulation can be used, where the level of QAM for all subcarriers is chosen based on measured parameters.

Let's consider this at the component level. The structure of a 4G transceiver is similar to any other wideband wireless transceiver. Variances from a typical transceiver are mainly in the baseband processing. A multicarrier-modulated signal appears to the RF/IF section of the transceiver as a broadband[4] high PAVR signal. Base stations and mobiles are distinguished in that base stations transmit and receive/decode more than one mobile, while a mobile is for a single user. A mobile may be a cell phone, a computer, or other personal communication device.

The line between RF and baseband will be closer for a 4G system. Data will be converted from analog to digital or vice versa at high data rates to increase the flexibility of the system. Also, typical RF components, such as power amplifiers and antennas, will require sophisticated signal processing techniques to create the capabilities needed for broadband high data rate signals.

Figure 19-1 shows a typical RF/IF section for a transceiver.[5] In the transmit path *inphase and quadrature* (I&Q), signals are upconverted to an IF, and then converted to RF and amplified for transmission. In the receive path, the data is taken from the antenna at RF, filtered, amplified, and downconverted for

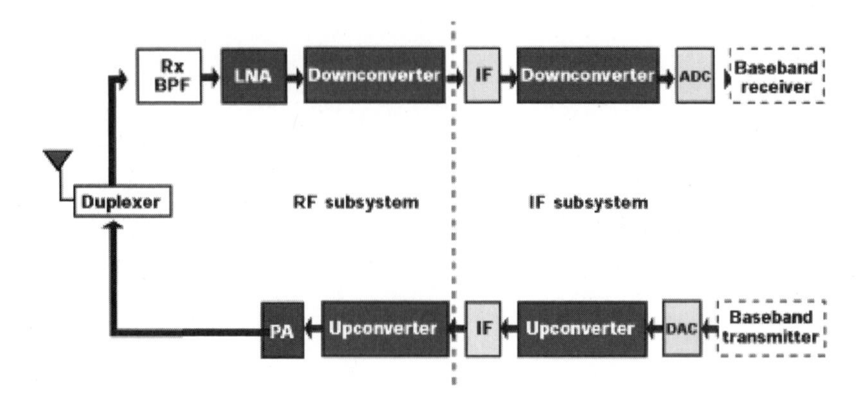

FIGURE 19-1 A basic RF/IF block diagram showing a typical RF/IF section for a transceiver.

baseband processing. The transceiver provides power control, timing, and synchronization, and frequency information. When multicarrier modulation is used, frequency information is crucial. If the data is not synchronized properly, the transceiver will not be able to decode it.

From a high level, the structure of the RF/IF portions of the mobile and base station are similar; however, significant differences are in their architectures and performance requirements. Key drivers for both are performance and cost; mobiles also need to consider power consumption and size.

4G Processing

Figure 19-2 shows a high-level block diagram of the transceiver baseband processing section.[6] Given that 4G is based on a multicarrier technique, key baseband components for the transmitter and receiver are the FFT and its inverse (IFFT). In the transmit path, the data is generated, coded, modulated, transformed, cyclically extended, and then passed to the RF/IF section. In the receive path, the cyclic extension is removed; the data is transformed, detected, and decoded. If the data is voice, it goes to a vocoder. The baseband subsystem will be implemented with a number of ICs, including *digital signal processors* (DSPs), microcontrollers, and ASICs. Software, an important

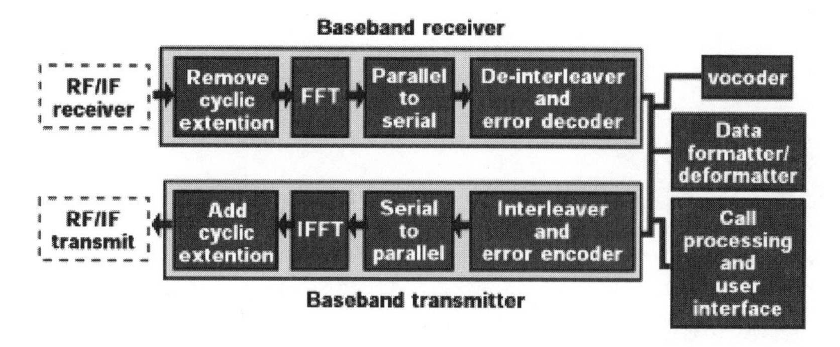

FIGURE 19-2 A high-level block diagram of the transceiver baseband processing section in a 4G wireless design.

part of the transceiver, implements the different algorithms, coding, and overall state machine of the transceiver. The base station could have numerous DSPs. For example, if smart antennas are used, each user needs access to a DSP to perform the needed adjustments to the antenna beam.

RECEIVER SECTION

4G will require an improved receiver section, compared to 3G, to achieve the desired performance in data rates and reliability of communication. For 3G using the 2-Mbps data rate in a 5-MHz bandwidth, the SNR is only 1.2 dB. In 4G, approximately 12-dB SNR is required for a 20-Mbps data rate in a 5-MHz bandwidth. This shows that for the increased data rates of 4G, the transceiver system must perform significantly better than 3G.

With any receiver, the main issues for efficiency and sensitivity are noise figure, gain, group delay, bandwidth, sensitivity, spurious rejection, and power consumption. 4G is no exception; the sensitivity can be determined.

For a 4G receiver using a 5-MHz RF bandwidth, 16 QAM modulation, and NF of 3 dB, the receiver sensitivity is –87 dBm. For 3G, the receiver sensitivity needs to be –122 dBm; the difference is due to the modulation and PAVR. This illustrates the need to reduce PAVR by clipping or coding. Also, the gain is required to be linear, and the group delay must be flat over the bandwidth of the signal.

The receiver front end provides a signal path from the antenna to the baseband processor. It consists of a bandpass filter, a *low-noise amplifier* (LNA), and a downconverter. Depending on the type of receiver, there could be two downconversions (as in a super-hetrodyne receiver), where one downconversion converts the signal to an IF. The signal is then filtered and then downconverted to or near the baseband to be sampled.

The other configuration has one downconversion, as in a homodyne (zero IF or ZIF) receiver, where the data is converted directly to baseband. The challenge in the receiver design is to achieve the required sensitivity, intermodulation, and spurious rejection, while operating at low power.

The First Line of Defense

The receiver bandpass filter is the first line of defense to eliminate unwanted interference and noise. This filter must be able to achieve the cutoff needed for each bandwidth. In a 4G implementation, the bandwidth could be as low as 5 MHz and as high as 20 MHz. If the filter were to be only 5-MHz wide, it would not have the capabilities to use the 20-MHz bandwidth. However, if the filter is 20-MHz wide and the signal is only 5-MHz wide, the extra interference would increase the noise and reduce sensitivity. This means that a tunable filter is needed. One option would be a bank of filters with different bandwidths, where selection is made based on the need.

A typical LNA has a noise figure of approximately 1 dB and a gain of about 20 dB. A trade-off is made between gain and noise to provide the best solution. The LNA sets the noise figure of the overall receiver because it is one of the first components of the receiver. Because of the high PAVR of the signal, the LNA will also have to be very linear to minimize any extra distortion.

The downconverter section of the receiver will have to achieve good linearity and noise figure while consuming minimal power. A measure of the linearity in the mixer section is the *spurious free dynamic range* (SFDR). This is directly related to the second and third order intermodulation products also known as IP2 and IP3.

The *analog-to-digital converter* (ADC) is the key component that can break the new system. System issues of the ADC concern whether or not to use undersampling, the PAVR of the signal, the bandwidth, and the sampling rate. For a 5-MHz bandwidth signal, a typical sampling rate would be 20 MHz. If IF sampling is used, the aperture uncertainty or jitter must be low enough to prevent errors.

The next requirement is the dynamic range. For an MCM system using the theoretical PAVR for a 512-point IFFT, the dynamic range required would be 80 dB, which is equal to 13 bits.

The desired quantization noise is determined by the average ratio of average signal power to average noise spectrum (see sidebar, "The Spectrum Wars") density measured in dB (Eb/No) for the subcarriers, the *data rate* (DR), and backoff (which is generally 15 dB). The constant 20 dB is added to the end to put the quantization noise 20 dB lower than the system noise.

If the signal has interference or blocking, the ADC requires additional bits. The required dynamic range of the ADC could increase from 15 to 17 bits.

BASEBAND PROCESSING

The error correction coding of 4G has not yet been proposed; however, it is known that 4G will provide different levels of *quality of service* (QoS), including data rates and bit error rates. It is likely that a form of concatenated coding will also be used, and this could be a turbo code as used in 3G, or a combination of a block code and a convolutional code. This increases the complexity of the baseband processing in the receive section.

4G baseband signal-processing components will include ASICs, DSPs, microcontrollers, and FPGAs. The receiver will take the data from the ADC, and then use it to detect the

THE SPECTRUM WARS

Mountain Telecommunications wants to bring wireless phone service to Indian reservations in the Southwest and other remote regions. So they petitioned the *Federal Communications Commission* (FCC) for access to an ideal spot on the airwaves. However, an unlikely rival has emerged as a competitor: The Pentagon hopes to run part of a national missile defense system on the same frequency.

A new generation of wireless telephones, satellite TVs, and Web-based personal planners is demanding prime space on the electromagnetic spectrum, the range of frequencies where radio and radar signals are carried. However, some of that real estate is already home to tenants like UHF television stations, which long ago obtained spectrum space to air snowy reruns of Hogan's Heroes. Now, with prime real estate going fast, even the military is having to vacate space to make room for the digital revolution.

Washington regulators are now trying to find a slice of spectrum for 3G cell phones. These so-called 3G devices are touted as powerful enough to carry streaming video and fast enough to juice e-commerce.[7] However, the Air Force and Navy use that bit of spectrum to fly their satellites, and NASA uses it to control the space shuttle. The FCC simply does not have enough spectrum to give everyone all that they want. The military has already given up large portions of spectrum since 1993.

The spectrum is afflicted by gadget gridlock. Cell phones, a luxury for some 340,000 U.S. subscribers in 1985, are a way of life for 300 million today. By 2003, 2 billion cell phones will be in use worldwide, half of them equipped to link to the Internet. Although a television station uses a slice of spectrum to send one signal to many viewers, each wireless phone user transmits on its own sliver of spectrum.

The military wants more space, too. Desert Storm, the first information-age war, was fought nearly a decade ago, eons in Internet years. Now a major report on the subject completed in 2000, "Joint Spectrum Vision 2010," finds that the Pentagon will need more access, not less, for commanders demanding real-time images and mobile and satellite communications.

Something has to give. For example, Mountain Telecommunications' plan to provide Indian reservations wireless access was rejected by the office that regulates government use of the spectrum. The Pentagon indicated sharing that piece of spectrum would have *a major negative impact on national security.* The FCC has yet to rule, but is likely to reject the bid.

WorldSpace, a provider of satellite digital audio services, stepped into this minefield in 1998. It planned to launch its AmeriStar satellite in 2000 to provide digital radio to the Caribbean and South America. However, the Pentagon indicated AmeriStar would interfere with frequencies set aside for aircraft and missile tests.

LINE IN THE SAND

WorldSpace is reconfiguring its satellite so that it will not interfere with Defense Department transmissions, at a cost of several million dollars. The Pentagon would prefer that no one else try to cross the line in the sand it has drawn. It has invested more than $100 billion in its spectrum infrastructure to support nearly 900,000 military systems. Reconfiguring weapons to function in different bands of spectrum would cost taxpayers billions.

It doesn't help that there is no national strategy for balancing security and commerce on the spectrum. A report by the Pentagon's Defense Science Board urged everyone to get along. The guiding light of the report is a plan that looks at how everyone might share spectrum. The military isn't interested. They are in their forts on the Potomac with their guns pointing out, ready to shoot anyone who wants part of their spectrum. Literally trillions of dollars are at stake here. Not to mention the nation's security.[8]

proper signals. Baseband processing techniques, such as smart antennas and multiuser detection, will be required to reduce interference.

MCM is a baseband process. The subcarriers are created using IFFT in the transmitter, and FFT is used in the receiver to recover the data. A fast DSP is needed for parsing and processing the data.

Different algorithms can be used to create a smart antenna; the goal is to improve the signal by adjusting the beam pattern of the antennas. The number of DSPs needed to implement a smart antenna depends on the type of algorithm used. The two basic types of smart antenna are switched-beam antennas and adaptive arrays. The former selects a beam pattern from a set of predetermined patterns, whereas the latter dynamically steers narrow beams toward multiple users. Generally speaking, SA is more likely be used in a base station than a mobile, due to size and power restrictions.

Multiuser detection (MUD) is used to eliminate the *multiple access interference* (MAI) present in CDMA systems. Based on the known spreading waveform for each user, MUD determines the signal from other users and can eliminate this from the desired signal. Mobile devices do not normally contain the spreading codes of the other users in the cell, so MUD will likely be implemented only in base stations, where it can improve the capacity of the reverse (mobile-to-base) link.

Transmitter Section

The purpose of the transmitter is to generate and send information. As the data rate for 4G increases, the need for a clean signal also increases. One way to increase capacity is to increase frequency reuse. As the cell size gets smaller to accommodate more frequency reuse, smaller base stations are required. Smaller cell sizes need less transmit power to reach the edge of the cell, though better system engineering is required to reduce intracell interference.

One critical issue to consider is spurious noise. The regulatory agencies have stringent requirements on the amount of unwanted noise that can be sent out of the range of the spectrum allocated. In addition, excess noise in the system can seriously diminish the system's capacity.

With the wider bandwidth system and high PAVR associated with 4G, it will be difficult to achieve good performance without help of linearity techniques (for example, predistortion of the signal to the PA). To effectively accomplish this task, feedback between the RF and baseband is required. The algorithm to perform the feedback is done in the DSP, which is part of the baseband data processing.

Power control will also be important in 4G to help achieve the desired performance; this helps in controlling high PAVR—different services need different levels of power due to the different rates and QoS levels required. Therefore, power control needs to be a very tight, closed loop. Baseband processing is just as critical whether dealing with the receiver or transmitter sections. As you've seen, RF and baseband work in tandem to produce 4G signals. The baseband processing of a 4G transmitter will obviously be more complicated than in a 3G design. Let's consider the chain of command.

The *digital-to-analog converter* (DAC) is an important piece of the transmit chain. It requires a high slew rate to minimize distortion, especially with the high PAVR of the MCM signals. Generally, data is oversampled 2.5 to 4 times; by increasing the oversampling ratio of the DAC, the step size between samples decreases. This minimizes distortion.

In the baseband processing section of the transmit chain, the signal is encoded, modulated, transformed using an IFFT, and then a cyclic extension is added. Dynamic packet assignment or dynamic frequency selection are techniques that can increase the capacity of the system. Feedback from the mobile is needed to accomplish these techniques. The baseband processing will have to be fast to support the high data rates.

Even as 3G begins to roll out, DoCoMo system designers and services providers are looking forward to a true wireless broadband cellular system, or 4G. To achieve the goals of 4G, technology will need to improve significantly in order to handle the intensive algorithms in the baseband processing and the wide bandwidth of a high PAVR signal. Novel techniques will also have to be employed to help the system achieve the desired capacity and throughput. High-performance signal processing will have to be used for the antenna systems, power amplifier, and detection of the signal.

Finally, wireless promises a bold future of mobile commerce and multimedia. Before we get there, the telecom industry will have to overcome five unspoken hurdles discussed next and brings this chapter to its conclusion.

CONCLUSION

This chapter presented an overview of the future of i-Mode. In conclusion, although Americans look to cable modems and DSL to fulfill their broadband fantasies, Europeans want bandwidth for their mobile phones. The continent's major telecom players have spent the last few years wandering in a financial desert, searching for an elusive oasis they are convinced will revive them. That oasis is the wireless Internet, specifically in the form of the multimedia and much-anticipated 3G mobile phones.

The persistence is understandable, not least because it's impossible to overstate the importance of wireless technology in Europe. At the start of 2001, 63 percent of *European Union* (EU) citizens had a mobile phone, more than twice the percentage who had home Internet access. In some countries (if you exclude, say, infants and prisoners), mobile penetration is more than 100 percent.

Like so many Net fantasies before it, 3G promises to run your life more smoothly than ever before and without wires. Here's how Nokia's (**http://www.nokia.com/main.html**) Web site describes it: *3G is videoconferencing in a taxi. 3G is watching clips from your favorite soap in the train; 3G is sending images straight from the field to headquarters for analysis; 3G is sharing your Moroccan vacation with your friend—from Morocco.*

WARNING

URLs are subject to change without notice.

It's also serious money. Mobile telecommunications in the EU is a $300 billion industry with a growth rate of 12.5 percent a year. Europe's various operators spent a staggering $240 billion in 2000 on licenses to offer 3G services and will likely spend twice that much in 2002 to build out their networks.

However, the 3G oasis as it currently is envisioned looks like a mirage. Financial analysts doubt the business proposition, engineers question the technology, and consumers emit devilishly mixed signals. Still, the telecoms continue to trudge through. There is no turning back—at this point, they have too much invested in 3G.

In the near future, the wireless Internet may play as important a role in our lives as the PC version does now. To get there, the wireless industry needs to disavow itself of some pretty overblown myths:

- **Myth 1** 3G technology will be so revolutionary that it will sell itself.
- **Myth 2** The success of NTT DoCoMo's i-Mode in Japan proves a lucrative audience exists for advanced wireless services.
- **Myth 3** Bluetooth will create a whole new world of devices that will stimulate the demand for wireless services, on and off the Internet. In just three to four years, Bluetooth will operate in more than 1 billion devices worldwide, letting computers talk to phones, PDAs, and regular appliances without any wires.
- **Myth 4** The wireless world, especially in Europe, is ready and eager to embrace m-commerce.
- **Myth 5** The imminent arrival of 3G phones will create a massive market for wireless data services.

MYTH 1

The promise of experiencing a true multimedia experience (most notably full-motion video) over a handheld device is seductive. That's what 3G technology has long promised: a bridge across the chasm between cell phones and PCs. Consumers know that cramming such rich capabilities into a cell phone is a tough task, and the mobile companies have admitted the challenge, all the while promising that the delays in bringing 3G to market lie mostly in the software. However, skeptics suspect the delay hints at a darker truth that 3G technology doesn't work.

Connection speeds for 3G will be nowhere near as fast as they need to be to deliver the data-intensive transmissions that have been promised. 3G operators have not thought about their battery problems. No solution is available yet.

The problem would be obvious to anyone who used a 3G phone: The device would become unbearably hot. A battery for a 3G phone operating at 2 Mbps would burn out every few minutes and would make the phone too hot to handle. Mobile engineers have admitted the problem and now project maximum speeds of 256 or 512 Kbps—which may not be enough to deliver the streaming video 3G companies have hyped about.

In late May 2001, when Japan's NTT DoCoMo gave away trial 3G handsets, analysts speculated that they operated at speeds lower than 200 Kbps. That was the least of DoCoMo's problems. Recently, the company recalled 2,700 handsets following customer complaints about the new 3G service, including reports that the phones generated excessive heat.

Even if 3G works flawlessly, it will be adopted at different times in different markets. This means that many consumers are going to need *hybrid* handsets that enable access both to 2.5G and 3G services, possibly for several years. There's only one problem: no one has yet seen such a device.

Myth 2

It's true that i-Mode has come much closer than any American or European service to providing a wireless Internet. Like the Internet, i-Mode breaks up and transmits information in discrete electronic packets, and i-Mode is compatible with HTML, so ordinary Web sites can be viewed from an i-Mode handset.

Perhaps most important, i-Mode has done what almost no Internet company has figured out: getting users to pay for content. According to DoCoMo, approximately 40 percent of the sites designed for i-Mode are premium (that is, paid) sites, and about half of i-Mode's 34 million users subscribe to at least one premium site.

That sounds impressive, until you look at what they pay: The fees are between $1 and $3 a month. You've got to have a very large customer base to make any money on that kind of pricing. It's going to be some time before the total U.S. or European 3G audience reaches 40 million. Assuming the same rate of premium subscribers, that means a maximum of about $38 million a month available to a few thousand content providers—not exactly an industry tidal wave.

Moreover, the business model that enables content sites to flourish through i-Mode depends on nearly all the revenue going back to the content providers. So far, U.S. and European mobile operators have been unwilling to share the wealth at that level. If they continue to balk, they may find themselves incapable of providing the personalized services that make i-Mode so popular. If they give over to i-Mode's business model, then it's not clear how they can make money. It's a riddle that few companies seem eager to solve.

MYTH 3

To many, Bluetooth represents wireless nirvana. Named after a tenth-century Viking king, Bluetooth was developed by two engineers from Sweden's Ericsson in 1994. Its basic function is to enable different electronic devices to connect with one another without wires. Your Palm (**http://www.palm.com/**) can talk to your computer; your computer can talk to the Internet without a phone line; and both can talk to your stereo.

For that to happen, though, the technology needs to prove it can work. So far, Bluetooth has suffered the quintessential technology horror story: trade show demonstrations that don't work. From California to Hanover, Germany, reporters and analysts have watched Bluetooth demonstrators fall flat on their faces. Bluetooth boosters insist such stories are purely anecdotal and that the bugs will be worked out.

It's true that Bluetooth is a powerful idea with some good science behind it. It uses the 2.4-GHz frequency, which, unlike the costly 3G spectrum, has the virtue of being unregulated. Both Compaq and Hewlett-Packard (**http://www.hp.com/**) recently released devices that are Bluetooth-compatible.

However, Bluetooth has some serious problems that have yet to be addressed. Start with compatibility: In the United States, wireless networking technologies already are in use. One of them, 802.11 (the horridly technical name comes from an engineering standard) is considered the market leader; Starbucks (**http://www.starbucks.com/Default.asp?cookie%5Ftest=1**) has announced it will install 802.11 in some of its U.S. cafes to enable customers with laptops wireless access to the Net. However, 802.11 is incompatible with Bluetooth because both use the same frequency spectrum. Companies are working to harmonize the two technologies, but as of yet, no firm has been able to solve the protocol problem. This technological conflict may be why Microsoft (**http://www.microsoft.com/**), an early proponent of Bluetooth, decided against incorporating Bluetooth into Windows XP.

Bluetooth also leaks like a rusty pail. Both Bluetooth and 802.11 appear to have significant security problems. Bluetooth's architecture is intrinsically insecure. A couple of high school kids with a scanner and a soldering iron could crack it in minutes. It wouldn't even slow down a professional.

Such predictions, like those of Y2K anarchy, may be bunk. However, until Bluetooth technology begins passing some crucial market tests, it's hard to see how it will be widely adopted for any real-world Internet applications.

Myth 4

Scores of companies have sprung up recently promising that in just a few years, Europeans will trade stocks, pay bills, and do a good deal of their shopping via their phones. Especially because mobile phones can locate customers precisely, the purveyors of m-commerce are confident that location-tailored telephone commerce is going to be huge.

It is certainly true that Europeans have agreed to pay for certain transactions over the phone, in some cases more quickly than could have been predicted. Approximately 20 billion SMS messages worldwide are sent every month, up from zero just a few years ago. Collectively, those text messages represent real money; some European telecom firms now count on SMS messages for between 10 and 15 percent of their revenue.

NOTE

In the United Kingdom alone, the number of SMS messages will pass 2 billion a month some time in 2002.

Consumers are billed for those services through their phones and at pennies per transaction. It's a huge leap from there to a Web-like system where the product or service is not delivered through the phone itself, and where the provider is separate from the merchant. Most mobile providers have no experience or infrastructure for the backend of m-commerce. The mobile guys have diddlysquat. They're going to have to form alliances with AOL and Microsoft, and that will eat into badly needed revenue.

Then there's the security issue again. Probably the least-discussed barrier to m-commerce is the lack of security and privacy[9] in wireless transactions. A recent survey conducted by Boston Consulting Group (**http://www.bcg.com/**) uncovered deep mistrust in this area. In Sweden, a country where mobile phones are as ubiquitous as anywhere on the planet (more than 8 out of 10 adults use them), an astounding 88 percent of consumers indicated they were concerned about sending their credit card number over a mobile network.

That surprisingly high figure could stem from the fact that, compared with many developed nations, Sweden has relatively low credit-card usage. Even so, the figure for concerned consumers in the United States is 75 percent; in Japan, it's 84 percent. Consumer fears about security leaks in Internet and wireless networks may be overblown. However, when a hacker working on a Brooklyn, N.Y., public library computer can steal banking and credit card details from the likes of George Soros and Ross Perot, as happened recently, public wariness seems justified. Overblown or not, consumer fear is a huge marketing challenge and one that mobile operators and m-commerce purveyors seem ill-equipped to tackle.

MYTH 5

This is the mighty myth, the one that launched $130 billion in frenzied license auctions. According to widespread predictions made in 1999, 3G phones were supposed to be on the market by now, offering customers mobile data speeds of up to 2 Mbps.

NOTE

That is approximately 40 times faster than a 56 Kbps dialup connection and would presumably allow for easy viewing of moving color images.

Instead, 2.5G phones are just now being introduced, and many companies have had to postpone their 3G launches. Siemens (**http://www.siemens.de/**) recently estimated that by middecade, only 15 percent of European subscribers will be using 3G.

By itself, delay does not mean failure. However, the pressure is acute because of the debt that major wireless operators acquired in order to obtain spectrum licenses.

NOTE

At Deutsche Telekom (**http://www.dtag.de/dtag/ipl2/cda/t1/**)*, for example, debt reached more than $50 billion, and the company's credit rating hinged on junk-bond status.*

Furthermore, the full rollout of 3G services requires building a vast network of radio towers and base stations. That expense will almost certainly add another $100 billion to the 3G bill, possibly twice that amount. Recently, the German government indicated it would enable 3G license winners to share network expenses. It's far from clear, though, that such post-auction niceties will withstand legal challenges from those who were outbid.

The longer these services are delayed, the more market share they may lose to cheaper services called 4G. There has not been one business plan for 3G that makes sense. Not one.

END NOTES

[1]*Making The Leap to 4G Wireless*, Motorola, Inc., Corporate Offices, 1303 East Algonquin Road, Schaumburg, IL 60196 USA, 2001.

[2]John R. Vacca, *The Cabling Handbook* (2nd Edition), Prentice Hall, Upper Saddle River, N.J., 2001.

[3]Eurotechnology Japan K.K., Parkwest Building 11th Floor, 6-12-1 Nishi-Shinjuku, Shinjuku-ku, Tokyo 160-0023, Japan, 2001.

[4]John R. Vacca, *Wireless Broadband Networks Handbook: 3G, LMDS, and Wireless Internet*, McGraw-Hill, New York, 2001.

[5]Motorola, Inc., Corporate Offices, 1303 East Algonquin Road, Schaumburg, IL 60196 USA, 2001.

[6]Ibid.

[7]John R. Vacca, *Electronic Commerce: Online Ordering and Digital Money with CD-ROM*, Charles River Media, 2001.

[8]WorldSpace Corporation (Headquarters), 2400 N Street, Washington, DC 20037 USA, 2001.

[9]John R. Vacca, *Net Privacy: A Guide to Developing and Implementing an Ironclad Business Privacy Plan*, McGraw-Hill, New York, 2001.

CHAPTER TWENTY

FUTURE MARKETS FOR I-MODE AND ITS HANDSET SUPPLIERS

Following extensive discussions on mobile data, the delivery of nonvoice information to a mobile device, analysts have been remarking in each of the past several years that the method-of-the-moment will usher in the new era of go-anywhere, do-anything communications. Notwithstanding those predictions, however, the latest estimates show that nowadays only 3 percent of all mobile traffic is data.

Contrary to technology, optimists gushing about the possibilities of instant-response, Web browsing, and full-motion video cell phones face an unsatisfied demand for voice services driving the growth of new subscribers. Others argue that, as with wireline Internet service, a critical mass must be achieved in order for widespread market appeal to materialize.

KEY FINDINGS

Meanwhile, a series of announcements from independent sources confirm the slowdown in subscriber growth and handset unit expansion worldwide. Citing recent statistics, subscriber penetration in Europe is quickly approaching an *inflexion point* of 55 percent, beyond which further expansion is set to be increasingly restrained, as is presently confirmed by Scandinavia. The growth rate is set to weaken to around $70 million from the current pace of $30 million, which represents a slowdown amounting to no less than 30 percent.

The global handset industry is also in a phase of deceleration —several of the largest producers of chipsets for wireless communications have lowered their output targets for the current year. Consensus on unit shipments in 2002 seems to be gradually contracting to the range of $600 to $657 million.

As such, global output is set to increase by 55 percent, compared with around 65 percent in 1999. In a further confirmation of the ongoing slowdown, leading cellular operators, such as BT Cellnet and Vodafone, are allegedly trimming their capital expenditure plans in *third generation* (3G) equipment deals with the chosen suppliers. This should hardly come as a surprise because they have to act carefully to compensate for the hefty license fees.

The opposing camp of analysts believe, however, that the mobile data sector has turned a corner in 2001, and that it is finally beginning to witness the long-promised growth in demand and revenues to mobile telephone providers and other wireless operators. Among the number of developments contributing to this change, in agreement with the *Federal Communications Commission* (FCC), existing mobile telephone subscribers provide a large potential market of mobile data users.

In some markets, wireline data traffic is almost equal to wireline voice traffic. A Yankee Group study shows that 57 percent of mobile telephone subscribers have Internet access at home. Therefore, it is logical to conclude that wireless carriers would be anxious to capture even a small portion of such a vast market.

According to Merrill Lynch, other potential growth drivers of the mobile data sector include the tremendous increase in digital handset use, the low retail price for *short messaging service* (SMS), and greater computer literacy. Some even dare to predict tremendous potential for growth of wireless data services in the United States.

One forecast estimates that by 2002, mobile data services will outnumber wireline data subscribers. Other analysts expect at least $46 to $51 billion in revenues by 2007, representing an annual growth rate of 25 to 30 percent, along with as much as 200 million subscribers using some form of wireless data.

3G Progression Services

To date, mobile telephone operators in the United States have begun to deploy 2.5G technologies, the first phase of the transition to 3G as described by some analysts. Equipment manufacturers are in the process of developing 3G equipment and operators are testing technologies.

Although some industry insiders observe that future 3G global networks might turn out to be fragmented and incompatible, in particular, lacking in roaming (one of the most critical features of wireless communications), the current mobile data capabilities of digital cell phone technologies and software, together with infrastructure upgrades, are available for each technology.

Several CDMA operators have worked out potential upgrades as they transition to 3G networks. One such upgrade is IS-95B, a packet-based network that combines eight channels and is expected to offer data transmission rates up to 64 Kbps. At the same time, foreign carriers have begun to deploy software upgrades in the core and radio networks for IS-95B at a cost of $50 to $120 million per operator. Analysts point at SK Telecom in Korea and DDI/IDO corporations in Japan as a good example of IS-95B deployment.

No carrier in the United States has announced its intent to adopt the upgrade thus far. Qualcomm is developing a data-only upgrade to work with CDMA, called *high data rate* (HDR). It offers rates up to 2.4 Mbps for fixed applications using the existing 1.25-MHz channel. U.S. West, which was acquired by Qwest Communications in June 2001, and Cisco Systems have begun a trial of HDR. Korean Telecom Freetel plans to adopt HDR in 2002.

Meanwhile, Sprint PCS (PCS) has started testing 1xRTT, a radio transmission technology that is often referred to as cdma2000 phase one, and allegedly plans initial deployment by mid-2002. Australian Telstra (TLS) began 1xRTT trials in June, while SK Telecom announced plans to introduce the service by the end of 2002.

Lucent Technologies and Sprint PCS announced a simultaneous trial of 1xEvolution technology (1xEV), which is based on HDR and incorporates the 1xRTT technology. The *CDMA Development Group* (CDG) claims that 1xEV enable data capabilities of at least 2 Mbps and even greater voice capacity in the existing 1.25-MHz channel.

In March 2001, Nokia and Motorola unveiled the joint development of 1XTREME technology that promises integrated voice and data, with transmission speeds of up to 5.2 Mbps. Also in the existing 1.25-MHz channel, 3xRTT, or cdma2000 phase two, supports packet data rates of up to 2 Mbps and higher, requires 5 MHz of spectrum. As its name implies, 3xRTT triples the bandwidth capabilities and requires three of CDMA's 1.25-MHz channels.

A number of carriers held tests during 2001. Analysts indicate that after 1xRTT, the choice between the higher speed networks may depend on a carrier's capital expenditure plans, spectrum considerations, and demands for high-speed data services by customers.

The other CDMA-based 3G standard is *wideband CDMA* (W-CDMA), which many European carriers are expected to deploy. It requires 5 MHz of bandwidth. Since it aroused a great deal of interest after the first three months of 2001, a couple of contracts have been executed and, based on experts' reports, several trials are in progress. These include efforts by Nortel Networks and Vodafone, as well as Ericsson and Telia in Sweden.

W-CDMA achieved most of its popularity after Japan's NTT DoCoMo commenced development of the technology, setting service objectives of data exchange at a maximum speed of 2 Mbps. The company is firmly determined to launch its W-CDMA service no later than May 2002.

Having deployed a *cellular digital packet data* (CPDP) overlay over their networks, TDMA carriers are currently working on the implementation of *general packet radio service* (GPRS), because it is compatible with both TDMA and *Global System for*

Mobile Communications (GSM) networks. GPRS is a packet-based, data-only upgrade that ultimately enables data rates up to 115 Kbps. In order to migrate to an interim 2.5G technology called *enhanced data for GSM evolution* (EDGE), TDMA carriers have to use a core GPRS backbone for IP and EDGE, as the radio interface technology. AT&T recently announced plans to use EDGE for its 3G strategy, while *SBC Communications* (SBC) and *BellSouth* (BLS) indicated that they would launch EDGE in late 2002 or early 2003.

At present, Lucent and Ericsson are running trials with AT&T. In the United States and abroad, GPRS is believed to be the primary migration path for GSM operators due to its reliance on packet rather than circuit switching, a factor of considerable importance. European operators estimate that the upgrade from GSM to GPRS will cost between $60 to $230 million. Among U.S. carriers, Omnipoint, prior to its merger with VoiceStream, expected the upgrade to GPRS to cost less than $10 million.

Although vendors have negotiated a number of GPRS contracts with both U.S. operators and those overseas, handsets are not yet commercially available. Industry insiders report that the merged VoiceStream/Omnipoint/Aerial plans to launch services by the end of 2002, and Powertel (PTEL) and PacBell Wireless are expected to begin deployment within the same terms.

FOR THE IMPORTANCE OF I-MODE

Much of what is driving the potential growth in U.S. mobile data services is based on trends observed elsewhere. As previously mentioned in other chapters, NTT DoCoMo's i-Mode service was launched in 1999 and had 34 million mobile data customers as of this writing. As a result, the company is currently the largest *Internet Service Provider* (ISP) in Japan, offering access to over 8,000 Web sites and content provided by more than 600 companies.

Being a simpler solution than *Wireless Application Protocol* (WAP) and capable of working on *second-generation* (2G) networks, i-Mode may be easier to bring to market. DoCoMo is deemed one of the most important companies to watch on the battleground for the mobile Internet, and its huge success in Japan gives it a technological and marketing edge over any global competitors.

However, Asia's shine in the wireless space is not only because of the i-Mode service. Recent statistics from DataQuest highlight the growing impact of leading Japanese manufacturers. Taking into consideration their stagnating domestic markets and the huge opportunities opening up in 3G, Japan and Korea are set to build a stronger presence in the overseas markets.

Panasonic, NEC, Mitsubishi, and Kyocera have made significant advances. Samsung has fallen victim to the ban of subsidies in Korea. The Japanese rise presents a special interest because the handset market is stagnating, forcing the Island of the Rising Sun to focus its efforts on export opportunities with special emphasis on GSM and TDMA, which make up a combined 77 percent of the world market potential as estimated by Dresdner Kleinwort Benson.

Facts, such as Sony's fresh launch of its state-of-the-art cmd-Z5 GSM model in Europe, Panasonic's readiness to ship 3.4 million TDMA handsets through its contract with AT&T Wireless, and increasing proliferation of i-Mode are indicative that the playing field will shift further in favor of the East by the middle of the decade when 3G handsets begin to reach mass-market volumes.

Although the high-profile battle is going in Europe and the Americas, mounting threats are emerging from China, the largest growing market of all. Unconfirmed reports claim that domestic suppliers now command some 12 percent of the Chinese handset market, a 10-fold increase from the figure in 2000. Deriving political preference and benefiting from the vernacular citizenship status, local vendors are likely to keep altering the market dynamics materially. With the adoption of a common 3G standard, the W-CDMA, Asian handset makers are set to bolster their market positions well before the transition to 3G products.

However, a fight is going on as handsets duke it out for the Internet. *Personal digital assistants* (PDAs) are being boxed in a corner by the latest cell phones, as this next part of the chapter on i-Mode wireless access to the Internet explains.

I-MODE HANDSETS AND CELL PHONES IN A BRAWL

Proponents of PDAs indicate that it will be these hand-held units that tie the on-the-go executive of the future to the wireless Internet. Coming out strong is a whole new generation of Web-enabled PDAs (like i-Mode), made not only by the manufacturers that have dominated the market so far, but a host of new ones as well, some backed by the hulking giant (Microsoft) from Redmond, Washington. Looking just as strong, however, is the other big contender for providing the executive wireless Internet access: the next-generation cell phone.

The smaller screens and keyboards available on cell phones, and their somewhat lesser processing capabilities, might seem to be crippling handicaps. Yet these phones already enable users to access the Web and exchange messages at a fraction of what it costs to do so using PDAs. What's more, the geographic availability of PDA networks remains quite limited, next-generation cell phones promise ubiquitous coverage within a single family of standards.

In 1993, when John Scully, then chief executive officer of Apple Computer Inc., Cupertino, California, introduced the first PDA, the Apple Newton, he foresaw a world in which a simple mobile device would connect mobile users to vast networks of information. Today, PDAs like the Palm VII from Palm Inc., Santa Clara, California, and any of several Windows CE-based units, use mobile radio-messaging services, such as BellSouth's Mobitex or GoAmerica, to give users a foretaste of that prediction. However, the services' geographic coverage is not extensive, and their low data rates make it frustrating to browse the World Wide Web.

Nine years after the Newton's unveiling (and two years [1998] after its demise), PDA manufacturers are still looking to fulfill Scully's vision. So, they are hoping that new wireless networking technologies, such as Bluetooth and the WAP, will make their products the dominant means of surfing the Web wirelessly. According to the Framingham, Massachusetts-based market research firm International Data Corp., those units are exactly what most people will be using by 2002. However, wireless phone companies and handset manufacturers will offer services and products running on the same protocols—and in one case, in Japan, the national service provider already has scored a knock-out with a system (i-Mode) that takes a shortcut to the Internet.

TODAY'S PDAS

For its third try at creating an *operating system* (OS) for PDAs that would finally catch on with the public, Microsoft, in April 2000, introduced Windows CE 3.0, thereby causing PC suppliers to launch a new flotilla of devices to run it. Microsoft dubbed the new PDAs Pocket PCs and hoped they would be able to grab market share from Palm operating-system-based devices (see Figure 20-1).[1]

The Pocket PCs come bundled with Microsoft's Pocket Internet Explorer browser. Thus, they can put Web pages on a PDA's 95-mm, 240 × 320-pixel screen, which is about one-sixth the size and one-fourth the resolution of a 380-mm, 640 × 480 desktop monitor, that is, the most primitive PC display one can find.

NOTE

In the case of the Palm products, screen resolution, at 160 × 160 pixels, is even less, but the size is about the same as the Pocket PCs. Yet, despite the size reduction, color PDAs provides a reasonable rendition of a not-too-graphical Web page.

Microsoft now indicates that the next generation of mass-market personal computing equipment will be some form of small, Internet-connected mobile device. However, the embat-

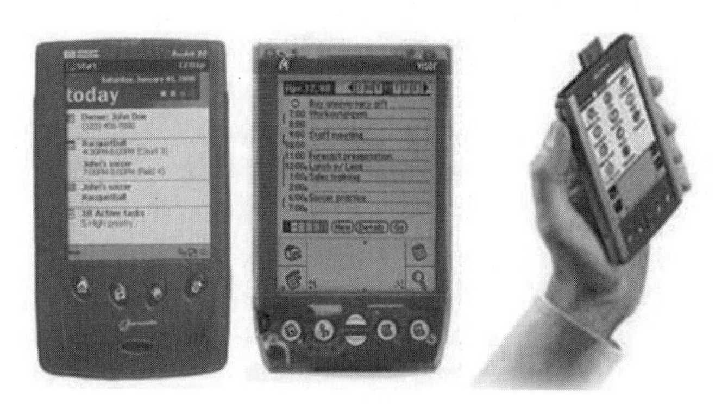

FIGURE 20-1 Internet-ready Pocket PCs like Hewlett-Packard Co.'s Jornada [left] combine calendars, MP3 audio, and the Web with color screens. However, Palm Inc.'s PDAs like the Visor [middle] own the market, so Sony Corp.'s new Net-ready PDA [right] relies on a Palm OS.

tled firm has, all the same, some way to go before it becomes a dominant factor in that arena. According to figures from market research firm PC Data Inc., Reston, Virginia, PDAs that use the Palm OS had a 95.8 percent of the PDA market in April 2000, leaving Windows CE with a mere 4.2 percent.

Microsoft hopes to leverage its PC-user base to duplicate its desktop success in the mobile market. Thus, the CE user-interface (on PDAs like Compaq's H3600, Hewlett-Packard's Jornada, and Casio's Cassiopeia) resembles that of desktop and laptop PCs. Microsoft's PDA applications, like Pocket Word for word processing and Pocket Excel for spreadsheets, mimic the functions of similar Microsoft Office programs (see Figure 20-2).[2] However, this attempt at mimicking the desktop on a handheld has not yet attracted many buyers.

The Palm OS market share is divided between Palm and Handspring Inc., Mountain View, California. The latter is the company started in 1998 by the developers of the original Palm device, the first commercially successful PDA. Although five of Palm's handhelds are sold for every two Handspring Visor PDAs, Handspring is scheduled in 2002 to introduce a new product. Industry observers think it will be the company's first color system, competing with the Palm V.

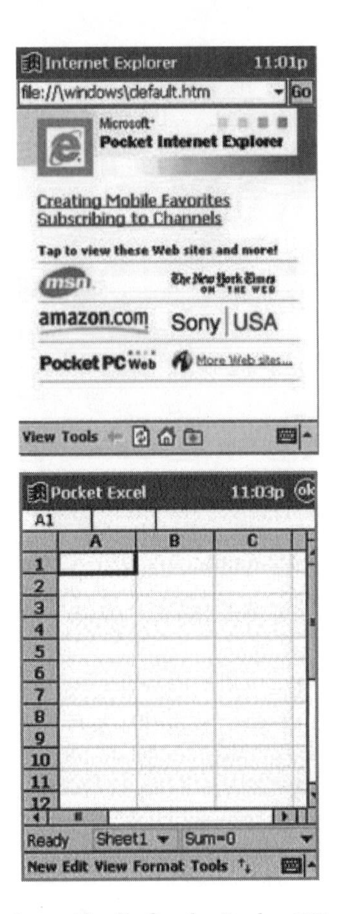

FIGURE 20-2 Designed specifically for the Pocket PC, the Internet Explorer software [top] is a pared-down version of Microsoft's Web browser. It lets the user do such things as entering a Web address and adding links to a list of favorite sites by clicking on the menu bar's folder icon at the screen's bottom. Similarly, Pocket Excel [in the following] is a Pocket PC-specific version of Microsoft's popular Excel spreadsheet.

Although all the new functions of the forthcoming Handspring device are not yet known, the ability to connect to a phone network is high on the list of desirable features. The bottom line here is that PDAs need to improve their ability to handle voice and mobile telephony if they are to succeed.

Efforts at providing wireless access to the Internet for the home might result eventually in new PDAs with voice capability also. This could be the case for Gateway Inc., North Sioux

City, South Dakota. Gateway and America Online Inc. (AOL), Dulles, Virginia, recently announced that they are jointly developing Internet appliances powered by the recently introduced Crusoe processor from Transmeta Corp., Santa Clara, California.

The new Net equipment would run Transmeta's Mobile Linux OS and the new Gecko Web browser developed by AOL and Netscape Communications Corp., Mountain View, California. The initial three devices (a wireless Web pad for use in the home with a base station, a countertop appliance for the kitchen or den, and a desktop appliance that is billed as *a lower-cost alternative to the traditional PC*) will be designed and built by Gateway. The desktop appliance will be on the shelves in time for holiday shopping in 2001, whereas the other two products will come to market in the first quarter of 2002. However, the company is mute on telephony capabilities for the new devices and its plans for taking the technology into the PDA arena.

LOOKING FOR SOLUTIONS

Recent advertisements would lead the public to believe that PDAs are the perfect way to get to the Net while on the go. In one ad, from wireless Internet service provider GoAmerica Communications Corp., Hackensack, New Jersey, a man in a three-piece suit is seen proudly holding up a PDA with an antenna; beneath him runs the ad copy: *Go to the whole wide Internet . . . for the right answers where you are.* So, aren't PDAs already the ideal wireless Web-surfing solution? Well, not really.

Although it is true that some PDAs can get to the Net wirelessly, they do not work everywhere—or even in most places outside urban areas. This drawback is in consequence of the limited coverage of their Internet-access backbone, which consists of RF networks provided by companies such as BellSouth Wireless Data LP, Woodbridge, New Jersey.

For instance, the Palm VII comes with a built-in wireless modem that lets it connect to Palm.net, a service the company provides for its wireless users. However, Palm.net's coverage is

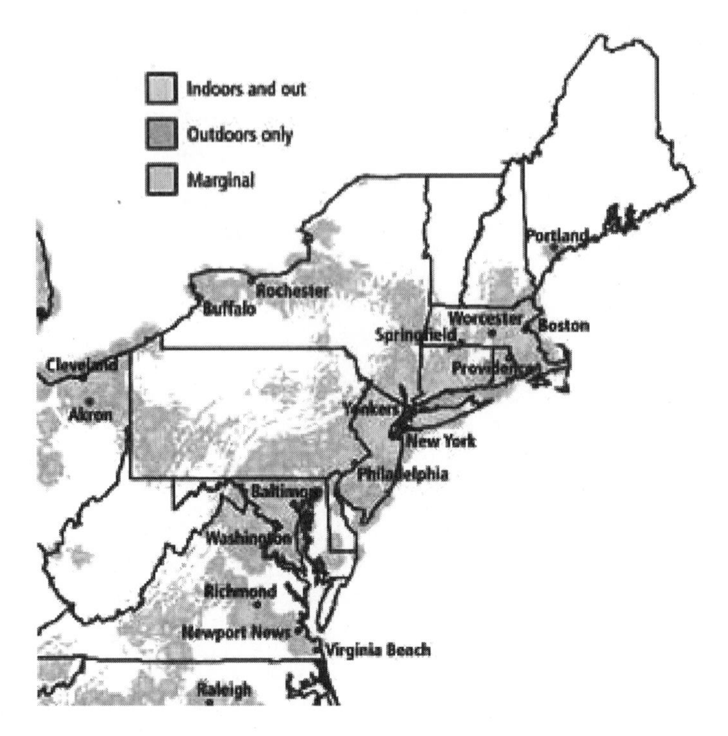

FIGURE 20-3 Areas in which PDAs can connect wirelessly to the Internet in the United States today are very limited. For instance, in the Northeast, which is one of the areas most heavily covered by BellSouth's Mobitex service, indoor and outdoor service is available only in large metropolitan areas. Even outdoor-only service drops off rapidly outside major suburbs.

spotty at best. Although Palms work well in major metropolitan areas, they have difficulty connecting beyond those boundaries.

So, for example, although Palm.net users can get e-mail easily indoors and out in New York City, coverage decreases as those people travel up the Hudson River. Just south of Albany in Hudson, New York, Palm.net is accessible only outdoors; north of Saratoga Springs, the network is unreachable (see Figure 20-3).[3]

This lack of service is due to Palm.net's dependence on BellSouth's Mobitex network for its RF connection. Mobitex was originally created in the '80s as a two-way text-messaging system for drivers and dispatchers in commercial taxi and limousine companies. The service's coverage is limited because demand outside highly populated areas has never been sufficient to justify the cost of installing transmitter/receivers.

It's a chicken-and-egg thing. Still, Palm is satisfied with the service for now.

At the time of the Palm VII's development in 1998, the company believed that it was not a practical option to build cell-phone circuitry into the device, so that it would have access to the Web through far more common radio network base stations. The cost and bulk of the additional cellular circuitry would have made the PDAs clumsy and cost prohibitive. Instead, Palm left it to third-party suppliers to provide links to separate cellular phones, which must be teamed with the PDA for a Net connection.

Nonetheless, being able to provide a wireless Net link was a top priority. Users were demanding, so Palm.net and Palm VII were born.

Palm is not the only company that saw this demand and chose special networks for Internet access. GoAmerica sells Internet access services for Windows CE PDAs as well as for Palm units. It, too, is mainly urban and lacks coverage in most rural areas of the United States. So far, neither Palm.net nor GoAmerica provide Internet access outside the United States, which is inconvenient for the globe-hopping executives who are a primary target for PDA marketers.

Another problem for BellSouth's Mobitex and other wireless RF networks is the slow rate at which they deliver Web pages. Users have complained about *the World Wide Wait* when using modems as fast as 56 Kbps. What's more, today the emphasis is on moving to the even higher data rates of digital subscriber lines and cable modems[4] for access to the Web. In contrast, the RF networks are working at rates of 9.6 Kbps—which is fine for plain text, but too slow for ordinary Web pages with any kind of pictures or graphics.

To tackle that problem, third parties currently offer so-called wireless solutions for PDAs using modems and cellular phones with higher data rates of 14.4 Kbps. However, they have one drawback: These solutions require wires. The user must connect the PDA to a cellular-phone with a special cable, an arrangement that is tough to juggle when walking down the street or around a conference center. Although it can work when the user is stationary (seated, say, in a restaurant), the need to be stationary is frowned upon by those looking for mobile communications.

Using PDAs for mobile Web applications is likely to be an uphill battle. If you are looking to surf the Web, you will be very, very unhappy.

Not Your Father's Phone

Certainly, PDAs have improved since the Newton was introduced, but so have telephones. Back then, phones were for the most part stationary. However, today cell phones are common, and the functions they perform go beyond simple conversation. Ericsson, Motorola, Nokia, and Kyocera Wireless offer so-called smart phones—phones that feature Internet access and other PDA-like features, such as the ability to store a calendar.

The Qualcomm QCP-1960 from Kyocera Wireless Corp., San Diego, California, which came onto the market in 1999, is typical of today's cell phones that can display Web pages. However, its screen, like those of other phones before it, is cramped, being limited to 4 lines of 12 characters and a single line of icons.

The phone relies on a customized version of the standard *hypertext mark-up language* (HTML), called the *hand-held device mark-up language* (HDML) (see Figure 20-4).[5] Like PDAs on RF networks, its 9.6-Kbps data rate is slow compared to desktop modems, but then the display does not need much data to fill it up, so it seems to work fast.

The 7110, the latest phone from Nokia Group, Espoo, Finland, is slated to hit the streets of Europe soon and will provide Internet access at 14.4 Kbps. The extra speed is warranted: The 7110 can display more information at a time on its 96 × 65-pixel display—large for a cell phone, but small compared to a PDA. The user scrolls the screen and selects links with a one-finger rolling button called a Navi (short for navigation) roller.

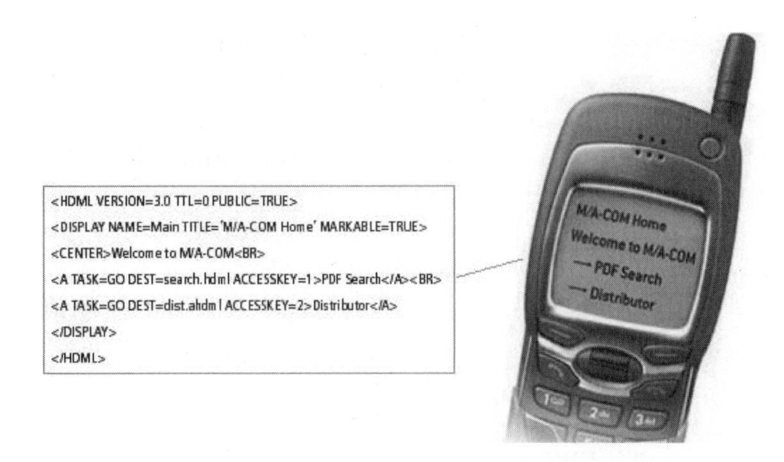
<HDML VERSION=3.0 TTL=0 PUBLIC=TRUE>
<DISPLAY NAME=Main TITLE='M/A-COM Home' MARKABLE=TRUE>
<CENTER>Welcome to M/A-COM
<A TASK=GO DEST=search.hdml ACCESSKEY=1>PDF Search</A>
<A TASK=GO DEST=dist.ahdml ACCESSKEY=2>Distributor</A>
</DISPLAY>
</HDML>
M/A-COM Home
Welcome to M/A-COM
PDF Search
Distributor

FIGURE 20-4 Information destined for the four-line screen of typical Web-enabled cell phones used on the streets of Europe and North America must be coded in the HDML [top], a custom version of the Web's HTML. Tags, such as ACCESSKEY=2, for example, indicates that pressing 2 on the keypad will select Distributor in the menu.

The software that puts the data on the screen is a microbrowser, a compact browser designed specifically to work with an industry-developed standard called WAP. Other phones, like Mitsubishi's T250, take the same approach.

The protocol has been in development since 1997 by the WAP Forum Ltd., Mountain View, California, which is made up of over 400 companies. Its members form a veritable Who's Who of the international telecommunications and computer industry: AT&T Wireless Services, Bell Atlantic, Compaq Computer, Deutsche Telekom Mobilnet, Ericsson, France Telecom, Hewlett-Packard, IBM, NTT, and more.

WAP was designed to deliver multimedia and digital services to mobile phones and other wireless devices once a wireless connection has been established. The standard defines the way devices should communicate, providing protocols that control everything from transport all the way up to the final application.

One of the things WAP defines is a new way of encoding Web pages for the small screens of mobile devices. Called the *wireless mark-up language* (WML), it is a successor to HDML, on which it is based. WML encodes information so a WAP browser (see Figure 20-5) can choose to display those things it is capable of handling.[6] For instance, for a simple display, the WAP browser will choose to display only the text portion of a page, although for a more complex screen, it will provide graphics. However, the WAP specification does not spell out which features of WML must be implemented in a WAP browser, leading to different WAP devices handling the same data differently.

However, as more phones support WAP, the feature set available on cell phones will increase. PDA favorites like electronic datebooks and address books are available on the Qualcomm phone, as well as on units from Ericsson and Nokia. These new devices are taking on more of the functions of PDAs and their screens are growing. In fact, Palm and Nokia are

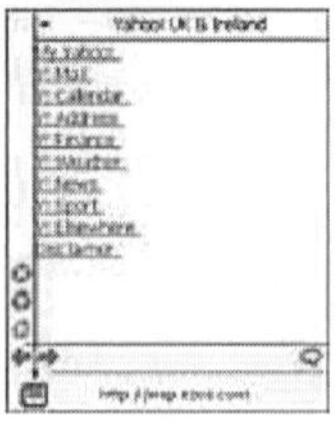

FIGURE 20-5 Flight reservations can be made on a cell phone (the two screens shown are in Swedish), thanks to WML. The same language can be tailored to the needs of the relatively larger display of a PDA in software like the EzWAP browser (below) from Ezos, Villeneuve d'Ascq, France.

working together to bring Palm's handwriting recognition technology to Nokia's cell phones.

Bluetooth to the Rescue?

In the spring of 2000, Palm, Handspring, Hewlett-Packard, and other manufacturers of PDAs indicated that they would use WAP and Bluetooth to wirelessly connect PDAs to Bluetooth-equipped cell phones and, through them, to the Net. Bluetooth is a technology for short-range data links.

The WAP-Bluetooth approach will make it possible for PDAs to work anywhere on the globe—the ultimate goal of mobile service providers. It looks like everybody's going in that direction.

Bluetooth employs radio frequencies in the 2.4-to-2.5-GHz ban—the so-called *industrial, scientific, and medical* (ISM) band, which is open for public use. Bluetooth-enabled devices will be able to communicate with each other over short distances; units that work at up to 100 meters apart are possible, but 10-meter or so applications are likely to be the most common.

Among the devices targeted by proponents of the technology (suppliers of both networking and computing equipment who are members of the Bluetooth *Special Interest Group* (SIG)) are PDAs, cell phones, computers, peripherals, and a host of other electronic gear. Bluetooth would let them all communicate directly with each other or through Bluetooth-enabled wired *local area networks* (LANs), perhaps across a meeting room or a football field.

Unlike IrDA, an infrared-transmission technology for small-area networks, Bluetooth does not require a line of sight to work and covers much greater distances. So a Bluetooth PDA could link to a Bluetooth cell phone or laptop, whereas the latter devices were sitting in a briefcase. Aside from providing Internet access by cell phone, Bluetooth would also let a user connect to a LAN while walking about an office thereby sending files to printers and e-mails to business associates. When it came up with the idea for Bluetooth in the mid-1990s, however,

Sweden's L.M. Ericsson had in mind letting two-cell phones talk to each other directly, without having to go through a phone company. Essentially, the Stockholm-based telecomm supplier wanted to combine a cell phone and a walkie-talkie.

Bluetooth's most impressive capability, though, seems to be its ability to gather support from industry. The technology came into the open in 1998 when Ericsson, IBM, Intel, Nokia, and Toshiba created the Bluetooth special interest group. After 3Com, Lucent Technologies, Microsoft, and Motorola joined the group in December 1999, the number of members climbed rapidly.

Today, the group has over 5,000 members and is still growing. The standards it originally worked out were adopted in June 2000 by the IEEE 802.15 working group for *wireless personal-area networks* (WPANs) as the foundation for a full-blown standard. The special interest group and the IEEE continue to work closely to formalize the standard, which should be released in the first quarter of 2002.

In advance of the formal IEEE document, Ericsson Mobile Communications unveiled the first Bluetooth cell phone, the T36 (see Figure 20-6) at a conference in Singapore, also in June 2000. The company will deliver the phone to wireless service providers in Europe during the second half of 2002.

The T36 phone's ability to stay active for long periods and move a lot of data makes it appropriate for Internet usage. The phone uses the *high-speed circuit-switched data* (HSCSD) format, the first step up in the rate at which GSM phones can send digital data on the road to the *Universal Mobile Telecommunication System* (UMTS).

Thus, it can communicate with cellular base stations at data rates up to 43.2 Kbps. The phone has 7.5 hours of talk time and 8.3 days of standby. Like Nokia's 7110 unit, it will also have a WAP browser.

There are good reasons for putting Bluetooth into cell phones. With Bluetooth, a cell phone could, say, wirelessly sync up with a PC to update an on-phone calendar or address book. Then, too, it could send information into a company's LAN. However, doing so now adds considerably to a phone's cost.

FIGURE 20-6 In June 2000, Ericsson introduced the first Bluetooth-compatible phone, the T36 [left]. A user wearing the Bluetooth-compatible wireless headset [inset] can press a button on the headset and talk on the phone while the latter stays put in a pocket or briefcase.

NOTE

Industry estimates of Bluetooth component costs alone are roughly U.S. $45 today. Consider that Bluetooth-less cell phones with browsers, such as the Qualcomm QCP-1960, sell for a mere $100. Adding Bluetooth hardware will increase the price of the phones significantly.

Not surprisingly, Bluetooth cell phones will initially be beyond the reach of most users. It is unlikely that Bluetooth functionality will be seen in entry-level or even midrange mobile handsets. Bluetooth will be offered as a high-end differentiator.

Nor will Palm build Bluetooth directly into its next Palm system. Initially, Bluetooth will be offered as an add-on module with an as-yet-undisclosed means of plugging into the PDA.

Most companies will not build the Bluetooth interface into Palm devices until the price of the parts falls to $5 per unit, which will probably not happen until sometime in 2002. So, what will it cost to connect a PDA to the Internet in 2002 when Bluetooth PDAs and phones reach retail shelves?

COST AND DURABILITY

Today, one can purchase a basic PDA for less than $200. Handspring's entry-level Visor, which has a monochrome display, lists for $149. However, for a color PDA like HP's Jornada, expect to pay $500 or more. Initially, Bluetooth modules for PDAs would cost about $250.

High-end phones today sell for $400 or more. If Bluetooth were to add just $45 to the cost of today's high-end cell phone, the least expensive monochrome Bluetooth PDA/cell-phone system would run approximately $850, versus $100 for an Internet cell phone like the QPC-1960. If color is added to the PDA, the price would shoot to around $1,200.

Even so, the cost for a PDA setup might be acceptable to well-heeled globetrotters if Bluetooth and WAP were going to be around for a while, becoming the standard means of mobile Web access. Although support being given Bluetooth by computer companies and the IEEE make it look like it will survive in the wireless personal networking arena, WAP's role in that area is not as assured.

According to a May 2000 report by British market researcher Ovum Ltd., London, a significant danger of disappointment and backlash is against WAP technology because of the delays in delivering it. After almost six years of work, the WAP Forum is still reviewing about a third of the documents that will go into making up the complete WAP standard.

However, the competition is not standing still. WAP and its WML are being challenged today in Europe by a technology already well established in Asia: I-Mode. As of this writing, the service had more than 34 million subscribers and demand was so great that NTT stopped advertising it for a while in order to boost network capacity and increase the supply of phones.

Key to this success is the fact that DoCoMo is delivering its services on handsets with displays like the 96 × 65-pixel screen of the Nokia 7110 mentioned earlier, higher resolution than most cell phones offer today. In fact, Nokia is a supplier of DoCoMo handsets, along with Matsushita Electric Industrial Co., Osaka, Japan, and Fujitsu Ltd., Tokyo; the latter is now supplying color-display phones.

Not content with its success at home, in May NTT DoCoMo, through Hutchison Telecommunications Ltd., Bristol, United Kingdom, a company in which DoCoMo has a 19 percent stake, launched its i-Mode service in Hong Kong. The same month, the Japanese wireless company agreed to invest 5 billion euros (U.S. $4.66 billion) in the Dutch wireless service provider KPN NV to establish a beachhead in Europe. This alliance will enable NTT to offer i-Mode as an alternative to WAP at a critical time. In July, the three firms announced that they would work together to launch an assault on Europe's 3G mobile phone market.

DoCoMo also has a working relationship with Telefónica in Brazil and owns a 7 percent stake in the Brazilian cell-phone service provider TeleSudeste Particiações. Then, too, it is eyeing the North American market; in November 1999, DoCoMo established a wholly owned subsidiary (NTT DoCoMo USA Inc.) in Palo Alto, California, and in April 2000, opened offices in New York City.

The company is now in talks with AT&T, SBC, and Verizon (the former Bell Atlantic Wireless). In late June 2000, DoCoMo announced that it would start putting English versions of its i-Mode portal on the Internet.

An advantage that i-Mode has over WAP is its reliance on ordinary HTML coding. I-Mode uses common HTML tags, so pages created for display on handsets can also be seen on PCs and vice versa. Using HTML makes it easy to create Web pages for i-Mode devices.

Indeed, many such pages can be found on the Web today. There were over 33,000 i-Mode sites in Japan as of this writing, and the number of sites was growing by 50 to 70 per day. Individuals and small businesses created many of the sites.

WAP, on the other hand, requires that WML, a new coding language, be used for WAP-compatible systems. So, WAP systems must come with special browsers capable of reading documents in that language. However, an advantage is that WML is also able to handle multimedia (video and audio), which the simple HTML that is used for i-Mode was not designed to do.

Although WML encoding can be produced automatically from existing Web pages, doing so is costly. IBM Corp.'s WebSphere Transcoding Publisher software, which works with its WebSphere server software, translates into WML existing Web pages encoded using HTML or XML, the *extended mark-up language* that is the latest form of Web mark-up language. The $20,000 software package is aimed at businesses running Web sites and at Internet service providers.

Likewise, software supplier AvantGo Inc., San Mateo, California, offers businesses a service that translates existing HTML pages to WML and provides guidelines on which HTML codes are supported by WML. For that service, a company must agree to share a portion of the revenue generated from the translated pages. How great a portion? AvantGo declines to disclose.

However, given the state of WAP's adoption today, even its supporters are hedging their bets. Ericsson's latest cell phone not only supports WML, it also handles ordinary HTML. The company is working with the U.S. discount brokerage firm (The Charles Schwab Corp., San Francisco) to develop mobile Internet applications for use in Hong Kong and Japan, markets already being served with HTML, as well as in Britain. Wireless phones are also starting to mature enough to accommodate the widespread adoption of Internet content.

CONCLUSION

This chapter discussed future i-Mode markets and its handset suppliers. In conclusion, turning to next-generation, off-the-wire communications again, the lack of a specific air interface compels users to face choices in mobile services as they do in today's cellular market. The variety of interim technologies adds to the confusion among operators.

Some see the solution in a trade-off, which not only will bring about simplicity, but will also enable consumers to have roaming capabilities on a fair number of networks around the world. The CDG, however, argues that dual-mode phones able to communicate over two types of networks, such as cdma2000 and W-CDMA, will offer the same kind of roaming many operators allow at present.

This may prove true once the whole world has migrated to fully formed W-CDMA or cdma2000 systems, which will most likely not happen smoothly because of the more complicated choices facing operators of the current two major types: GSM and TDMA.

This is particularly important for the United States. Although GPRS is essentially an upgrade of existing GSM networks, the implementation of UMTS, which is said to be a real 3G technology, will require entirely new radio equipment in both networks and user devices.

It makes sense for different operators to make such large investments when local market conditions justify it (at different times). It seems like the choice of technology is just the beginning of the struggle for operators, and in the longer term, for their customers.

Equally important is engineering and building the physical network. What determines its success is that for the 2G digital arena, the question has always been how the network is designed rather than which is the better technology.

For consumers' part, with the wireless industry moving forward in developing data service and marketing cell phones' abilities to provide e-mail and Internet access to consumers, some potential changes may be more bothersome, a survey from Hart Research Associates shows. A majority of all wireless users believe that technology drawbacks would outweigh benefits, especially in the case of having to carry a larger and heavier phone.

However, 25 percent of users express interest in future mobile phones with data services. Regarding the drawbacks of having to switch wireless ISPs and the additional costs of service, a majority of users feel that the benefits outweigh the drawbacks.

END NOTES

[1]Richard Comeford, "Handhelds Duke It Out for the Internet," IEEE Corporate Office, 3 Park Avenue, 17th Floor, New York, New York, 10016-5997 U.S.A., 2001.

[2]Ibid.

[3]Ibid.

[4]John R. Vacca, *The Cabling Handbook* (2nd Edition), Prentice Hall, Upper Saddle River, N.J., 2001.

[5]Richard Comeford, "Handhelds Duke It Out for the Internet," IEEE Corporate Office, 3 Park Avenue, 17th Floor, New York, New York, 10016-5997 U.S.A., 2001.

[6]Ibid.

CHAPTER TWENTY-ONE

SUMMARY, CONCLUSIONS, AND RECOMMENDATIONS

Since its launch only a little over two years ago, NTT DoCoMo's i-Mode service has grown to have more than 34 million customers, making it a benchmark for future cellular data provision, including *third generation* (3G). In this final chapter, a summary of i-Mode services is provided as well as future plans for the system.

SUMMARY

In general terms, what is i-Mode and how does it work? In addition to voice communications, i-Mode handsets also enable you to use e-mail and access the Internet. The i-Mode service enables users to access Internet services via their cell phones (see sidebar, "I-Mode Background").

DoCoMo's i-Mode connects users to a wide range of handy online services, many of which are interactive. These services include mobile banking, news and stock updates, telephone directory services, restaurant guides, ticket reservations, purchase of books and CDs, and much more. All services linked

I-MODE BACKGROUND

I-Mode is an Internet access system for cellular phones developed by NTT DoCoMo and launched by them on the market in February 1999. Over the past two years, its phenomenal growth has made it the standard-bearer for mobile data provision to public cellular customers.

The system works on top of the Japanese *Personal Digital Cellular* (PDC) system. Unlike *Wireless Application Protocol* (WAP), which modifies HTTP traffic for transmission, i-Mode passes the HTML (Web) pages straight through the system, although a subset of HTML is used. The HTTP and *secure HTTP* (HTTPS) layers are maintained, whereas the lower layers are processed at a gateway from TCP and IP on the Internet to packet mode for the radio link. Web pages can be up to 5KB in size, although 2KB is recommended, with 94 × 72 pixel GIFs. As well as Web traffic, i-Mode supports e-mail that the system pushes to the mobile phone, again using packet mode. E-mail can be up to 500 bytes (250 Japanese characters).

PDC is a TDMA system, standardized by the *Association of Radio Industries and Business* (ARIB), the Japanese standardization body. Each 25-kHz carrier can carry three full-rate or six half-rate speech channels. The physical layer of the PDC system uses (π4-QPSK with BCH codes for forward error correction. This maintains a bit error rate of about 10^{2} to 10^{4}. Operating on top of this is a link layer, *Link Access Procedure for Digital Mobile* (LAPD-M) that uses selective repeat ARQ *automatic repeat request* (ARQ) to maintain an error rate of 10^{6} or better. The TL layer uses stop go ARQ to give a 9.6Kbps packet data stream with packets of up to 1,400 bytes for the application layers.[1]

directly to the DoCoMo i-Mode portal Web site can be accessed immediately by simply pushing the cell phone's dedicated i-Mode button (see sidebar, "Into the Stratosphere by the Hundreds"). Users can also access thousands of voluntary sites via WWW addresses. Because i-Mode is based on packet data transmission technology, users are charged only for the amount of information they retrieve and not for how long they are online.

INTO THE STRATOSPHERE BY THE HUNDREDS

CargoLifter hopes to launch a fleet of 200 giant airships that will criss-cross the planet at an altitude of no more than 600 feet as shown in Figure 21-1.[2] The 250 giant airships that an American company called Sky Station International hopes to launch will fly much, much higher, around 65,000 feet, and without pilots. Technical details of the Sky Station plan are difficult to come by because the company, fearing competitors, refuses to discuss them. In fact, several groups right now are interested in putting a fleet of unmanned airships into the stratosphere.

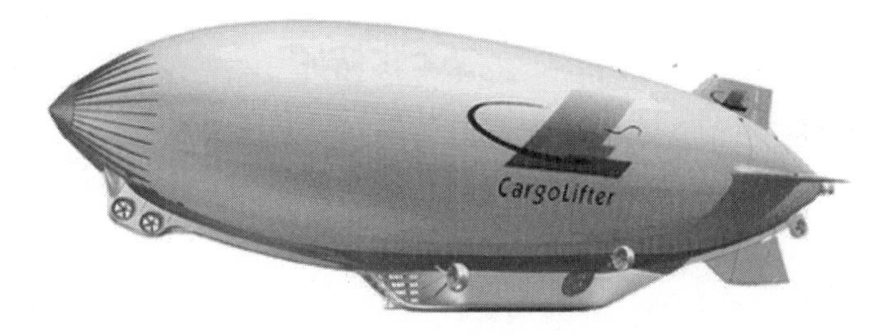

FIGURE 21-1 CargoLifter giant airship.

The idea is this: Cell phones are becoming ubiquitous; yet spotty reception makes them maddeningly unreliable. As more people start using their cell phones to connect to the Internet, and eventually to transmit large data files, the problems of inadequate capacity and intermittent coverage will become acute—even as the landscape becomes increasingly littered with ugly relay towers. However, talking by satellite requires expensive high-power phones that few people really want, whereas the costs of launching a global network are huge, as Iridium, the almost defunct satellite-phone company, discovered.

An unmanned airship floating in the stratosphere, 12 miles up, would not face the same problems. Ordinary cell phones would have enough power to send signals to it. Hovering above a city, it would add capacity and coverage—you probably wouldn't see it, but it would almost always be in your telephone's line of sight. In industrialized countries, airships could complement existing telephone networks. In underdeveloped countries, they could be even more important: With a

continues

continued

single airship and a lot of cell phones, you could provide phone service to a whole city that had none.

The biggest technical challenge is designing a stratospheric airship that can keep on-station. The rough design calls for a 720-foot airship whose top surface would be covered with thin-film solar cells made of amorphous silicon. Solar energy collected during the day would be stored in fuel cells for use at night and would power a single propeller, 100 feet in diameter, at the stern. Its huge size and slow rate of rotation would make it more efficient in the thin air of the stratosphere. The propeller would swivel like a rudder to help point the airship into the wind— which would be its nemesis. The airship could draw on other sources of power besides solar cells to achieve the speed it needs to stay on-station even in winter, when winds in the stratosphere can reach 110 mph.

The biggest challenge of all, however (as with all airship projects), is finding the money. The problem is to convince people that an old technology can be used for new purposes. Stratospheric airship proponents face a more conventional competitor: An American firm called Angel Technologies plans to use piloted airplanes instead of unmanned airships as high-altitude communications antennae (see sidebar, "Unmanned Planes as Telecom Towers"). The pilots would fly tight circles in the stratosphere above their service area for six hours, until the next plane arrived to take over. However, it is doubtful that many pilots would find that job fulfilling—and there is something curiously retro about the scheme. It seems a bit like proposing a return to human switchboard operators. Sometimes the most modern and forward-looking technology is not the one that fits our preconceptions; sometimes the elegant solution is a big, fat, slow-moving blimp.[3]

UNMANNED PLANES AS TELECOM TOWERS

As initially predicted in this author's 1999 book *Satellite Encryption* (Academic Press), NASA recently unveiled the world's first unmanned plane intended as a telecom tower in the sky, and it is attracting interest as a new way to get broadband Internet connections to businesses and to i-Mode and cell phone type users.

Helios, an aircraft resembling a giant wing, was built with funding and research help from NASA and has flown successfully. Backers claim its transmission services will be far cheaper than satellites and more efficient than wireless towers.

According to SkyTower Telecommunications, very poor broadband last-mile coverage is in the world, and they are looking to provide a wireless link to do it. SkyTower is a young subsidiary of solar-powered vehicles pioneer AeroVironment, which built and designed Helios. SkyTower hopes to begin mass production of the flying wings in 2003 and is in talks with potential partners.

SkyTower's foray into the commercial world is a breakthrough of sorts for scientists working on federal government-funded projects in Washington, D.C. Getting Helios to launch took $150 million of tax payers' money and the development of a new way for private companies to contract with the federal government. The project involved research by NASA and engineering by a consortium of private companies.

SkyTower is not unique in looking to the skies for an answer to the broadband bottleneck. Angel Technologies has a plan to use a light airplane designed to fly in the stratosphere; this is its *High Altitude Long Operation* (HALO) Network. The airplane has tested successfully, but no customers have been announced yet. Platforms Wireless International wants to use blimps for wireless communications at an altitude of 15,000 feet and has signed one customer—Americel of Brazil.

However, Helios is unique in its design and in plans for its use. With a wingspan of 247 feet (wider than a Boeing 747), Helios is 6 feet high and weighs 1,850 pounds, which enables it to take off at just 30 mph. It flies on the edge of Earth's atmosphere, 100,000 feet high. Helios' 14 electric motors run on solar power generated by 65,000 solar cells by day and on fuel cells energized by solar power by night. Helios' *brain* is an Apple Computer Macintosh computer that would guide it back to Earth when necessary.

continues

continued

Helios will be able to stay in the air for six months or longer because of its fuel cells and a limited number of moving parts. At an anticipated cost of $10 million each, it will be far cheaper than conventional communications satellites, which cost about $200 million each.

That is why Helios will soon develop into a platform of choice for fixed broadband, next-generation wireless, narrowband and direct broadcast applications. Helios can supply data rates of 1.5 to 125 *milliseconds* (ms) for a single user. The 30-ms latency of Helios-centered communications is comparable to that of fiber optics.

The plan is to launch the craft over large metropolitan areas such as New York City and San Francisco. The planes would essentially fly in circles, providing nonstop transmission for broadband services.

SkyTower's plan is to work with existing service providers to offer new services. They will be manufacturing and operating aircraft, as well as procuring and supervising development of payloads and customer premises equipment.

Backers are seeking investors to finance construction of Helios aircraft and partners among telecom companies that would develop services using the new platform. Analysts tracking flying platforms indicate the price tag of such solutions will be the main factor in their success or failure.

Flying planes such as this gets a little bit expensive. Cheaper solutions exist, like blimps or flying platforms. SkyTower has a chance of raising money if they can prove the costs are offset by the coverage they are going to offer.

Technical experts at the telecom providers that SkyTower is soliciting as partners have not yet formed an opinion about the viability of flying platforms such as Helios. The Helios technology doesn't seem ready for prime time.

Helios-powered broadband service would require customers to buy or lease a satellite dish and a router hub. They would subscribe to Internet access as a service.

USER APPLICATIONS THAT I-MODE CURRENTLY SUPPORTS

I-Mode mobile phones are equipped with 9600 bps packet-communication capability and browser software that can read a subset of HTML. Java-compatible 503 series i-Mode terminals have the following main functions:

- Support of permanent memory devices
- Support of multimedia data
- Automatic booting of an application
- Enhanced security[4]

Moreover, they support two versions of *Secure Sockets Layer* (SSL), SSL v2 and SSL v3, for enhanced security. This is a data security technology that prevents the interception of data, the falsification of personal identities and tampering with existing data.

WHAT I-MODE WILL SUPPORT IN THE FUTURE

DoCoMo intends to enhance its i-Mode services by adding additional banking functions and forging ties with providers of content such as Sony's Playstation. The forthcoming introduction of an advanced wireless network[5] based on *Wideband CDMA* (W-CDMA) technology prompted by NTT DoCoMo will provide an all-new platform for accessing increased Web content and applications via i-Mode, that is also provided in the current PDC system in the 50× series.

NOTE

The most popular i-Mode service is entertainment sites.

Creating New Application or Web Pages for I-Mode

How hard is it to develop applications and design Web pages with i-Mode? It is very easy because i-Mode supports programming in a subset of HTML, a de facto description language for Web pages. On DoCoMo's site, they also show you how to design i-Mode Web pages (**http://www.nttdocomo.com/i/tagindex.html**). Today, more than 50,000 sites are for the i-Mode service.

I-Mode Mobile Commerce Support and Security During Electronic Transactions

Twelve companies, including NTT DoCoMo have recently announced plans to jointly promote the *Edy* prepaid electronic money service and the use of *Edy* cards in the mobile environment as of December 25, 2000. As for current services, use of the SSL protocol for mobile communications and the issue of client certifications have further improved the level of security of Java-compatible terminals.

NOTE

I-Mode has 34,286,923 subscribers as of this writing.

NOTE

I-Mode has the following content providers: official sites of 2,600 as of this writing and nonofficial sites of 51,370, as of this writing.

Charging for Service

I-Mode users are charged according to the volume of data they transmit and not for the length of time they are online or the distance over which the data is transmitted. The basic packet transmission charge is equal to 0.3 yen per packet (128 bytes). Therefore, short e-mail of about 20 full characters can be sent for as little as 1 yen (about a third of a cent), a lengthier e-mail of 250 characters would be about 4 yen, and an airline reservation may be sent for as little as 20 yen ($0.07). I-Mode users also pay DoCoMo a 300 yen ($1) monthly charge in addition to the standard monthly charge for voice service. Additional information charges are payable to content providers when subscribers use certain i-Mode sites on DoCoMo's portal site for information.

NOTE

Content provider fees are billed through DoCoMo, which receives a commission (9 percent of information charges) from providers for its billing and collection services.

NOTE

I-Mode is available throughout Japan. The service can be used at high speed and even in the bullet train, which travels at an average speed of 200 kilometers per hour.

So, what are DoCoMo's goals for i-Mode development?

DEVELOPMENT

DoCoMo developed i-Mode technology so users can exchange information anytime, anywhere, and with anyone using i-Mode handsets. They will continue to listen to their users by developing additional services and content based on what users tell DoCoMo they need.

NOTE

Is there an increased use of i-Mode with Personal Digital Assistants (PDAs) rather than mobile phones? That is difficult to answer because i-Mode handsets and PDAs tend to be used for different purposes. It is indicated that more cell phones are being shipped and more cell phone users are worldwide.

JAVA'S AVAILABILITY AND SERVICES ON I-MODE PHONES On January 18, 2001, NTT DoCoMo announced the launch of a new i-Mode service based on Sun Microsystems' Java technology. The service dubbed *iappli* officially started on January 26, 2001. Java technology will enable users to run a wider variety of programs, including advanced video games, and process information with greater security, compared to current i-Mode technology.

Providers will be able to offer a wide variety of contents, and DoCoMo will continue to see much more content produced specifically for the i-Mode service. Users will be able to access high-quality content, which will result in wider use of the service.

NOTE

I-Mode can work with other air interfaces (GPRS, for example). It is possible to use the i-Mode service as long as handsets support i-Mode-compliant HTML.

NOTE

DoCoMo will continue to offer the i-Mode service. By exploiting FOMA's high-performance characteristics, DoCoMo will provide a more sophisticated i-Mode service.

Relationship to WAP

NTT DoCoMo is a major player in the WAP forum. I-Mode and WAP are presently competing technologies, but in the longer term there will be a convergence. In other words, i-Mode and WAP will eventually be integrated into X-HTML.

I-Mode has been very successful in comparison to WAP. Does the reason have to do with technology or marketing? It is a combination of both. Technically, the i-Mode service supports HTML, an Internet language, so anyone can produce content. It also uses the packet communications network that enables for volume-based charges, which have made the service available at low cost. In terms of marketing, DoCoMo uses easy-to-understand terminology, and they classify content into four portfolios, such as e-commerce,[6] information, database, and entertainment, to facilitate easy access to information and to encourage further use.

The bottom line here is that i-Mode's success is transferring to other markets like the United States and Europe. Cell phone usage is spreading worldwide. At the same time, cell phones are becoming smaller and more portable. By enabling these devices to process e-mail and access the Internet, as well as facilitate the instant exchange of information, DoCoMo is enabling themselves to participate in the new information society.

NOTE

HTML is a de facto Internet language that enables anyone to become an information provider. That is why DoCoMo decided to adopt it.

CONCLUSIONS

Web sites have historically been grounded in the meta data system language known as HTML, which isn't compatible with a WAP phone. WAP phones are best served by wireless markup language.

Unless a Web site is written in WML, a WAP phone can't access it. Only 35,000 WAP-accessible sites are in the world, according to wireless resource Pinpoint.com.

However, practically every major telecommunications company worldwide is using WAP as the de facto standard to transmit data on cell phones. Most users would be more inclined to use the Web capability on their phones if U.S. phone companies would adopt a service similar to that of Japanese telecommunications giant NTT DoCoMo's i-Mode, which can read practically any Web page (with varying degrees of legibility) and charges users for the amount of information downloaded rather than air time.

I-Mode is served by *compact HTML*, or cHTML, which technically can enable users to access desktop HTML sites, although it looks better if it's been written in cHTML. Because WAP defines a new markup language, content providers have to learn how to make content with it. Of course, this reasoning is one side of what is becoming one of the most contentious wireless Web debates.

Those who agree with the previous reasoning indicate that Japan's i-Mode offers more affordable access rates, more robust content, and higher connection speed. Nevertheless, through the WAP Forum, WAP boasts the worldwide support of over 500 major phone carriers and manufacturers that have been working together to ensure that their services are compatible with each other. The forum was founded in 1996 by Unwired Planet (now Phone.com).

NTT DoCoMo's i-Mode, meanwhile, is a proprietary service only offered in Japan and can't be made readily available on any other service carrier's network. There is a lot of industry support, and a lot of phones ship with WAP.

No statistics are available on the number of WAP phones in operation or the number of WAP subscribers worldwide because industry officials and analysts haven't counted. However, market analyst Datamonitor estimates that 27.6 million WAP subscribers are in Europe.

In Japan, there are over 34 million i-Mode subscribers and a little over 4 million WAP users, according to NTT DoCoMo. Officials from NTT DoCoMo and the WAP Forum adamantly deny the two technologies compete with each other. In fact, NTT DoCoMo is a member of the WAP Forum. However, the two groups still take jabs at each other.

WAP is put together by consensus. It's a democratic process by 600-plus companies around the world.

I-Mode is a specification. If i-Mode were a superior specification or technology, then other companies would have adopted it by now. However, 600-and-some companies have gotten behind the WAP standard rather than the i-Mode standard. That's got to tell you something.

WAP and i-Mode users may agree on one thing: Retrieving Web content on the i-Mode is much easier. Before accessing a site, WAP users must agree to pay extra charges and even type in URLs to browse through sites other than the service provider's portal. I-Mode phones have a one-button browsing method, eliminating the need to type in Web addresses.

I-Mode is more simple, but more people are using WAP because it covers so many countries. Also, the transfer speed for i-Mode is fast, and WAP is slow at this moment.

The i-Mode offers a lot of content for a fairly inexpensive price. You can send messages relatively inexpensively: for 100 characters, about a cent and a half.

Analysts may love i-Mode, but they admit that NTT DoCoMo is going to have a tough time extending its service past Japanese borders because of WAP's worldwide dominance. Because i-Mode is a proprietary, closed standard, American companies, European companies, and some Asian companies want an open standard where they can have a choice of vendors. WAP is designed to work on any network platform. The technology of i-Mode cannot be deployed on other networks.

Still, to the delight of mobile phone users, NTT DoCoMo has indicated it will offer i-Mode services abroad. Recently, NTT DoCoMo has taken a 15-percent stake in Dutch KPN

Mobile and claimed 20 percent of Hutchison 3G (the alliance between KPN DoCoMo and Hutchison Whampoa) to bid for 3G licenses in Europe.

Rumors are circulating that the company is poised to purchase a 10- to 15-percent stake in the joint venture between BellSouth and SBC Communications. The company is already on its way to completing the purchase of an Internet service provider: Verio Communications. That deal moved ahead recently after the White House shrugged off concerns that the United States could be left vulnerable to espionage if Japan's government-owned NTT gained access to U.S. wiretapping activities.

Despite the WAP Forum's belief that its protocol will dominate the future of the mobile wireless Web, i-Mode lovers aren't so sure. Companies like Microsoft are bracing themselves to support both WAP and i-Mode.

Right now, this is a young market and many of the standards and market positions are still evolving. Both formats have qualities that the other could benefit from. That's why Microsoft supports both HTML and WAP with Microsoft Mobile Explorer.

RECOMMENDATIONS

This final part of this final chapter is a direct response to common challenges raised by *chief information officers* (CIOs) worldwide. It is a comprehensive look at the concerns your peers have raised and will help shed light on the issues you face in addressing and making recommendations for mobile computing. The recommendations cover topics like

- Application mobilization
- Controlling communications and support costs
- Managing and supporting mobile devices
- How to cut through the wireless hype

- Lowering TCO of mobile devices
- Understanding the big picture[7]

What follows are the top 10 recommendations organizations are implementing today to build competitive advantage and deal with increasingly critical mobile and wireless computing issues.

ONE: DEVELOP A MOBILE STRATEGY NOW

Increasing market pressures coupled with the rapid-fire growth of mobile computing has created a booming population of mobile and remote workers. The drive to stay competitive has tasked today's enterprise with exploiting any and every means necessary in optimizing service levels, increasing sales, boosting efficiency, and cutting costs. Mobile computing provides the enterprise with several compelling competitive advantages, including

- Faster, decentralized decision-making
- Increased responsiveness to customers
- Increased sensitivity to market changes
- Lowered commuting costs/time for staff
- Increased staff morale and productivity
- Reduced travel costs company-wide
- Decreased facilities costs[8]

Enterprise demand for support of mobile computing initiatives now requires extending the full complement of enterprise resources to do business anywhere, at any time. The enterprise that proactively pursues a comprehensive mobile computing strategy will be successful in building competitive advantage. True business agility requires flexible technologies and the ubiquitous proliferation of computing power.

Failure to architect and build a mobile strategy today will have the same effect as ignoring the invasion of PCs back in the '80s. By developing a mobile strategy, which includes adopting standards, developing mobile infrastructure, and embracing mobile devices, your enterprise can effectively use mobile computing to stay competitive.

TWO: KEEP AN EYE ON WIRELESS

Wireless holds much promise for mobile computing. From real-time access for mission-critical applications to automated dissemination of competitive information, wireless will dramatically affect the mobile computing landscape. However, mobile and wireless are not interchangeable terms. Wireless is one component of mobile. Though wireless has substantial potential and some interesting uses today, a myriad of applications may prove more useable via wireline connections. Wireless computing is a tricky endeavor, with numerous pitfalls ready to snare the enterprise that moves without careful consideration.

WIRELESS TODAY Today's wireless networks are characterized by competing standards and protocols. No single network technology or operator will meet all your wireless network needs. Most of the current wireless data networks (known as 2G networks: GSM, CDPD, Mobitext, Motient) are built on analog and cellular digital networks—an infrastructure designed to support voice communications. Data transmission is a more complex endeavor, and the current public networks are ill-suited to efficiently support acceptable wireless data transmission rates. 3G networks are better equipped to handle data transmission, but are not slated for completion for years to come. Figure 21-2 cites a Yankee Group study into enterprise concerns regarding wireless adoption.[9]

PERFORMANCE As of this writing, practical wireless data speeds seldom eclipse 14.4 Kbps. Because users are typically accustomed to quick wireline connections, the allure of connecting wirelessly slackens in the face of limited bandwidth.

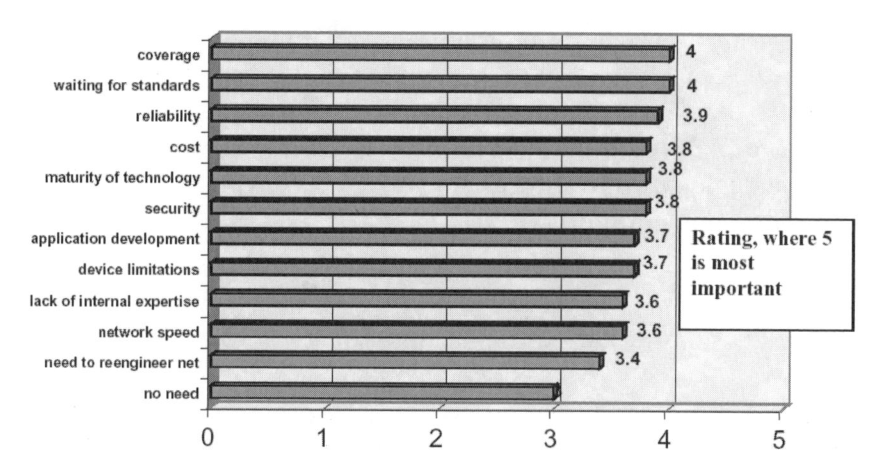

FIGURE 21-2 Barriers to wireless adoption.

RELIABILITY *Second Generation* (2G) wireless data networks are built on HTTP and TCP/IP-based protocols. Because HTTP and TCP/IP are particularly chatty packet-based technologies, they are better served by a high-speed wireline connection protected from interference and safeguarded against dropped signals.

COVERAGE Wireless coverage is spotty at best. Due to the lack of standard wireless protocols, coverage is governed by regional infrastructure. The most widespread wireless networks, like Mobitex and Motient, leverage existing cellular and digital infrastructure at the expense of speed (typically around 2 Kbps).

LACK OF STANDARDS 2G networks are clogged with a number of competing proprietary protocols. 3G seeks to alleviate this problem by offering one universal protocol.

WIRELESS TOMORROW Predictions on the future of wireless run the gamut from total ubiquity with the advent of 3G networks, to incomplete worldwide infrastructure build-out and lack of standardization for years to come. The reality is most likely somewhere in between. According to MobileInfo.com, 3G networks, when fully implemented, will move mobile computing to a new level after five years. Meanwhile, wireless applications should be implemented carefully.

3G networks aspire to be better stewards of available bandwidth while offering constant connections. Despite its potential, 3G currently faces only passive interest in the United States, with Europe beginning to invest the necessary billions to build out this next-generation infrastructure. Meanwhile, a host of other companies are trying to build new proprietary networks that offer a compromise between what is available today and what will be available in the distant future with 3G. Indeed, mobile wireless technologies will remain in almost constant influx until 2006/2007.

Wireless is an exciting technology that will change the face of mobile computing. It is important to understand how wireless fits into your business model so that your infrastructure is capable of supporting wireless as it exists today and as it will exist tomorrow.

Three: Accommodate the Occasionally Connected User—Real-Time or Synchronization

The fundamental challenge in mobilizing your enterprise is determining how a variety of mobile device users interact with data and information currently located on company servers. In addressing the challenge, the enterprise must decide between two scenarios:

- **Real-Time access** Real-time environments offer wireline or wireless access to a variety of enterprise applications. Users are dependent on a network connection to interact with the necessary information.
- **Synchronization** Also referred to as store-and-forward, synchronization enables users to work offline, connecting to the network only occasionally to sync up.

With real-time access, mobile devices are essentially viewers of server data, maintaining no local data stores. Because synchronization stores some data locally, users are not required to maintain constant connections to servers.

THE CASE FOR SYNCHRONIZATION Although constant, real-time access is appealing and well-suited for certain functions, synchronization may often be the more practical, smarter solution for the majority of enterprise needs. Synchronization offers the following benefits over real-time access:

- Reduced queries and network traffic
- Reduced user idle time
- Compression of staged data
- Reduced concurrent server processing loads
- Controlled communication costs[10]

As wireless protocols mature, the lines between real-time and store-and-forward architectures will begin to blur, and organizations will deploy both options in a complementary fashion. According to the Gartner Group, the convergence of synchronization and real-time mechanisms is crucial in accommodating varying bandwidth and connection scenarios and in graceful switching between modes of operation. The goal of 24/7 network availability can be shattered by the unreliability of a dial-up connection, and remote access to applications could prove useless when software is not in sync with desktop PCs.

The best way to accommodate the reality of occasionally connected users is to build a flexible infrastructure. Whether connected to corporate networks in a real-time environment, or working with localized applications in a deferred access environment, the optimal solution is to afford end users the luxury of being indifferent to, if not unaware of, whether or not they are connected. Users must avoid strategic investments in transitioning mobile wireless technologies and focus instead on developing backend logic that is device/network-agnostic and developing expertise in mobile application usability.

The current state of wireless technology, coupled with the inconvenience of staying perpetually connected via wireline, has created the reality of the occasionally connected user. Your enterprise should build a mobile infrastructure flexible enough to support both real-time access and synchronization.

FOUR: DEPLOY E-MAIL AND PIM TO HANDHELDS

Since its debut, e-mail's tenure as the most killer enterprise application has remained relatively unchallenged. E-mail is the application wireless adopters are asking for the most. As e-mail continues to be a vital method of communication, the ability to synchronize anywhere at any time and have access to the corporate intranet will provide significant productivity gains.

The advent of robust groupware applications, like Microsoft Exchange and Lotus Notes, has complemented enterprise e-mail systems with *Personal Information Management* (PIM) data, including calendars, contacts, to-do lists, and memos—providing one unified package for corporate workers to manage their busy lives.

Mobilizing groupware applications to handhelds is more complex than a first glance would suggest. Early solutions featured a handheld-to-desktop synchronization model. With these products, the full burden of installation, support, and troubleshooting rested entirely on the shoulders of those least likely to be able to perform these functions—end users.

Because these applications are not server based, the flood of mobile devices through the corporate backdoor is further complicated. A new breed of e-mail and PIM sync solutions now enables the exchange of data directly between handhelds and more functional server-based groupware applications. Important features of a server-based e-mail and PIM sync solution include

- One-step synchronization
- Connection transparency for users
- Complex filtering
- Encryption
- Flexible conflict resolution[11]

Enabling anywhere access to e-mail and groupware servers is the critical first step in empowering mobile workforces. Mobile workers are immediately more productive, and the sense of disconnect associated with being away from the office is minimized.

FIVE: PLAN FOR MULTIPLE DEVICES

If your organization is like most, there is already a mix of laptops, Palm devices, and Pocket PCs in use by staff and probably purchased by the company if only via expense reports. This is creating challenges for most IT shops. These range from network/data integrity issues, to an overtaxed helpdesk fielding support calls on devices it may or may not know exist, to inefficient systems management tools that don't work well for occasionally connected devices.

It was hard enough just trying to support laptops; then handhelds invaded your company through the back door. As the popularity of these devices continues to grow, and the variety of models increases, the feature lists on these devices continue to grow, as does the threat they pose to your corporation's information integrity as well as the costs, skills, and time required to support these devices. The time to take action is now.

DEVICE DIVERSITY One needs to only look at the success of Palm and other handheld PDA manufacturers to gauge the blossoming proliferation of mobile devices used as companion devices to the venerable laptop. That flood of mobile devices is only going to continue: by 2004, each corporate knowledge worker will have three to four different computing and information access devices that will be used to access various applications.

Although Palm's market share in the consumer handheld space has remained dominant, various device manufacturers are gaining ground—especially in the enterprise market. According to IDC, By 2004, Microsoft's market share will surge to 40 percent, compared with 51 percent for Palm.

With the popularity of handhelds reaching a fevered pitch, it's easy to forget about the laptop. For most, the laptop is the mobile workhorse. The effects of the recent slackening in growth of the PC market have not been mirrored in the laptop and notebook PC market. According to the Gartner Group, the worldwide mobile PC market grew by 32.7 percent in the third quarter of 2000 compared with the third quarter of 1999.

The promise of Bluetooth and the growing popularity of wireless LANs will further encourage enterprise adoption of laptops over the coming years. Many experts predict that over 50 percent of PC shipments will be laptop or notebook units within a few years. Falling prices and improved features, coupled with the increased market pressures forcing enterprise mobilization contribute to the adoption of mobile PCs and handhelds instead of traditional desktop workstations.

One of the most important points to take from this discussion is the notion that multiple device proliferation will continue, as Figure 21-3 suggests.[12] In a study conducted by Yankelovich Partners, a high technology research firm, when given the choice of carrying a wireless phone, two-way e-mail device, PDA, or pager, only 34 percent of professionals indicated they still wanted only one device.

You should not be bound by selective mobile infrastructure solutions that exclude certain devices. Business drivers should determine which devices are appropriate for which user groups, not your mobile platform software. In this way, you maximize business value and return.

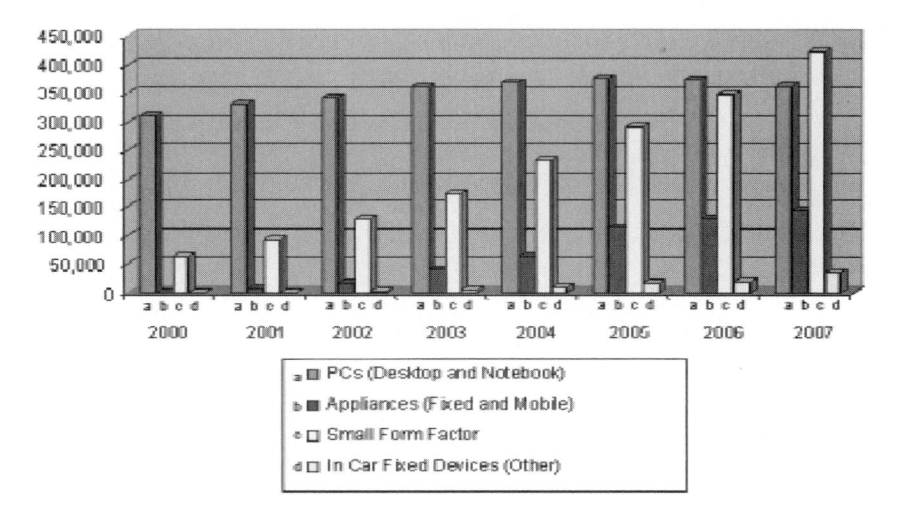

FIGURE 21-3 Total computing device usage—world forecast (in 000s).

SIX: STRUCTURE AND AUTOMATE CONTENT DISTRIBUTION

Getting access to file-based content and intranet pages is another area that can be challenging to mobile workers. Automated content distribution makes things much easier for the end user, enabling access to unstructured data including, spreadsheets, word processing documents, presentations, and graphics files. Ideally, this information is replicated throughout the mobile network. Without any effort, users have access to important files when on the road—even if a network connection is unavailable.

Although data synchronization helps proliferate structured data, typically a wealth of information is unstructured and saved in a variety of popular file formats. Many companies rely on e-mail to publish these files to end users. This methodology unnecessarily exposes the enterprise to several risks, including

- **E-mail viruses** No longer relying on e-mail attachments to move files through your network reduces the risk of exposure to viruses.

- **File versioning pitfalls** Users don't have to sift through e-mails to find the latest version of a file.
- **Mailbox administration** As attachments are less necessary, the stress on groupware servers due to large mailbox sizes is reduced.[13]

With automated content distribution, the most current files are automatically maintained and delivered to the appropriate personnel with no user intervention required. Content distribution also enhances the effectiveness of your corporate intranet. By making the site available offline as well as online, your intranet becomes a more effective, relied-upon communication tool. Components of a robust content distribution mechanism include

- Publish and subscribe architecture
- Web publication
- Remote device backup
- Overwrite versus rename
- File differencing
- Delivery logging
- Subscription management
- File versioning[14]

NOTE

Through implementing a content distribution protocol, your mobile users get access to the most current, time sensitive information found in files, including management reports, operations statements, pricing and product info, contracts, company forms and policies, and competitive information.

SEVEN: IMPLEMENT A ROBUST ASSET MANAGEMENT SOLUTION

Critical to the mobilization of your network data is the deployment of a solution that will enable your IT staff to remotely manage device hardware and software inventories. Gathering devices or burning CDs can be expensive and wastes valuable resources when remote software distribution can automate software installs, upgrades, and removals. Likewise, fielding support calls from mobile staff without an image of their device is extremely difficult.

Traditional LAN-based asset management solutions fail in the reality of the occasionally connected user. These solutions presume high bandwidth, always connected devices, with high network reliability. The reality for mobile workers is different—they connect their devices with the network only occasionally and typically over low bandwidth connections and frequently dropped connections. Traditional systems management vendors have been slow to support mobile users, and as a result, these capabilities are typically sourced from the new breed of mobile infrastructure solutions vendors. Of course, integration with the existing systems management solution is important here.

The systems management application you select should be comprehensive and flexible, providing customizable tools for systems maintenance, support, and troubleshooting. Must-haves in an asset management solution include

- Comprehensive user profiling
- Condition-triggered alert mechanisms
- Flexible real-time logging
- Hierarchical log construction
- Console-based log views
- Encryption
- Full-device refresh
- Checkpoint restart

- Transaction rollback
- File compression
- Default user Profiling
- Offline synchronization[15]

NOTE

With a powerful systems management solution in place, you are poised to deliver top quality support to end users, avoid undue strains on your scarce IT resources, and lower the total cost of ownership of mobile devices.

EIGHT: MOBILIZE APPLICATIONS THROUGH SYNCHRONIZATION

For the enterprise dedicated to building competitive advantage through extending the reach of its network, the mobilization of core enterprise applications is of utmost importance. The following applications can yield significant benefits through mobilization for your field-based and frequently traveling workforces:

- Sales force automation
- Customer relationship management
- Enterprise resource planning
- Field service applications
- Supply chain management
- e-business applications.[16]

An advanced data synchronization package, capable of supporting mobile PCs and handhelds alike, is the only way to ensure mobile workers have constant access to critical corporate data. Careful consideration should be given to selection of data synchronization technology. Your solution should fit inside your existing applications elegantly and cleanly, freeing your technical staff from writing complex conduit code and extensive integration effort. Make sure that the synchronization logic you define can be leveraged across multiple devices, and that you are not saddled with different administrative tools to support PCs, handhelds, and other devices. Important things to look for in data synchronization tools include

- Multiple platform support
- Multiple database support
- Open application development
- Field-level synchronization
- Offline synchronization
- Flexible change capture
- Graphical rules wizard
- Store-and-forward architecture
- Flexible conflict resolution tools
- Nonintrusive to applications[17]

NOTE

With an advanced data synchronization engine, you will be able to easily mobilize your core enterprise applications without extensive integration and conduit coding—making mobile workers more productive by enabling them to do business anywhere.

Nine: Beware Consumer-Focused Vendors

In evaluating partners to help architect and build your mobile strategy, be sure your vendors have demonstrated enterprise experience. Many *enterprise* solutions are repackaged consumer solutions and lack the features required for success in the enterprise environment. Your solution should be built from the ground up with the enterprise in mind.

Security Enterprise mobilization is inherently susceptible to security breaches. One reason some enterprises have delayed *going mobile* is security. Though their hesitance is understandable, it should not be a deterrent. Handheld devices are already carrying valuable enterprise data out the door everyday. As previously mentioned, the more functional they become, the more data is left uncontrolled. Everyday your corporation goes without a firm mobile strategy in place, the integrity of your network is jeopardized.

Your infrastructure should incorporate industry standard security techniques to encrypt the connection between devices and servers. For example, Certicom's Elliptic Curve Cryptography is an excellent tool to protect handheld data during communications sessions. Open APIs should be available to enable more advanced security features if desired. Direct integration of other security-related products and processes is made available via published APIs to ensure maximum flexibility for this sensitive topic. For mobile PCs, a relatively straightforward solution is to use a Web server utilizing SSL to secure communication with the laptop or tablet device.

Scaleable Look for a vendor with significant enterprise experience that is willing to take the time to understand your environment and assure you of system scalability. Be sure to consider not only the number of users you will need to support, but also the sync patterns that determine the number of concurrent sync sessions.

The mobile communications server must offer a clustered architecture to ensure basic scalability. In addition, dynamic load balancing reduces the stress on your servers by elegantly allocating users to different servers based on capacity. This architecture stands in stark contrast to static systems where users are married to specific servers, often resulting in unnecessary hardware purchases. Failover and recovery ensure that when a server fails during a transaction, it is transparently switched to alternate servers. The failed transaction is recovered and applied correctly by the new server.

ADMINISTRATIVE CONTROL Your solution should contain flexible administrative controls that enable your mobile infrastructure to change as dictated by business dynamics. At a minimum, your administrative console should enable you to

- Define the user base.
- Define synchronization activities.
- Set default configurations for sessions.
- Configure the amount of user control allowed.
- Subscribe users to activities.
- Prioritize the order of activities.
- Review extensive system logs.
- Set alerts and notifications.
- Remotely troubleshoot and address problems.[18]

NOTE

Make sure your mobile solutions partner has demonstrated experience in the enterprise market and can offer secure, scaleable, and flexible tools so your mobile infrastructure can grow with your business.

TEN: AVOID POINT SOLUTIONS

Finally, integral to the development and execution of a robust mobile strategy is a commitment to infrastructure development and extension. Your IT team should not waste time integrating piecemeal mobile solutions. Because infrastructure is the basic, enabling framework of the organization and its systems, a holistic approach to its design, deployment, and management is pivotal to organizational success.

A comprehensive mobile infrastructure, remotely managed through a well-equipped administrative console enables corporations to deploy groupware to handhelds, mobilize enterprise applications, control assets, and manage and deliver content. The alternative is a mix of incompatible point solutions with proprietary systems management and support consoles—overtaxing enterprise resources and jeopardizing network integrity. Your infrastructure should be capable of supporting a variety of devices and platforms including

- Laptop and tablet PCs
- Remote desktop PCs
- Palm *operating system* (OS) devices
- Windows CE/Pocket PC devices
- Industrial handheld devices
- Point-of-sales systems
- Barcode Readers
- Portable data terminals[19]

Figure 21-4 shows the components of a mobile infrastructure robust enough to support all of the mobile initiatives previously mentioned. When considering mobile infrastructure products, map their solutions against the model in Figure 21-4.[20] Even if you don't require all of this functionality today, building an infrastructure capable of supporting these functions will help you avoid the pitfalls of point solutions and recognize the long-term benefits of a comprehensive infrastructure solution. These benefits include

- Lowered training costs for administrators
- Easier for users—less effort to *get all their stuff*
- Flexible and easy for administrators
- Decreased support costs
- Decreased integration costs

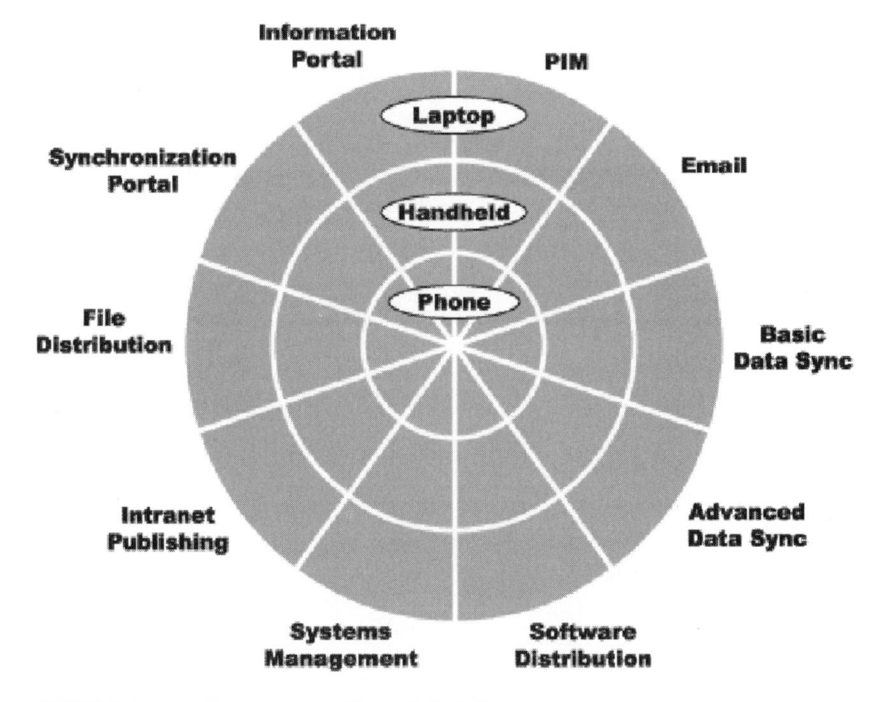

FIGURE 21-4 Components of a mobile infratsructure.

- Support for all your devices
- Lowered software license costs
- Less time evaluating/negotiating vendors
- One-point of contact for support/troubleshooting
- Increased user and administrator productivity[21]

Today's CIO is being asked to architect a mobile enterprise and extend the reach of the business. This author hopes this book has provided you with the set of recommendations that you need to support mobile computing and i-Mode and shed light on the complex swirl of issues surrounding mobilizing your enterprise—anywhere, anytime.

END NOTES

[1]Kei-ichi Enolci, "I-Mode," IEEE Corporate Office, 3 Park Avenue, 17th Floor, New York, New York, 10016-5997 U.S.A., 2001.

[2]Robert Kunzig, "Dirigibles on the Rise," CargoLifter AG, Potsdamer Platz 10, D-10785 Berlin, Germany, 2002.

[3]Ibid.

[4]IEEE Corporate Office, 3 Park Avenue, 17th Floor, New York, New York, 10016-5997 U.S.A., 2001.

[5]John R. Vacca, *Wireless Broadband Networks Handbook: 3G, LMDS and Wireless Internet*, McGraw-Hill, New York, 2001.

[6]John R. Vacca, *Electronic Commerce: Online Ordering and Digital Money with CD-ROM*, Charles River Media, 2001.

[7]"CIO Outlook 2001: Architecting Mobility," Synchronologic, Inc., 200 North Point Center East, Suite 600, Alpharetta, GA, 30022, 2001.

[8]Ibid.

[9]Ibid.

[10]Ibid.

[11]Ibid.

[12]Ibid.

[13]Ibid.

[14]Ibid.

[15]Ibid.

[16]Ibid.

[17]Ibid.

[18]Ibid.

[19]Ibid.

[20]Ibid.

[21]Ibid.

APPENDIX A

LIST OF I-MODE RESOURCES

The following tables (Tables A-1 through A-4) contain i-Mode and mobile technology resources on the Web. Some of the existing material is in Japanese, although an increasing amount is becoming available in English. Stay tuned for more.

TABLE A-1 I-Mode Resources

Resource	Description
Blink.com	Blink enables users to upload, add, and manage bookmarks in a free account accessible online or through a WAP or i-Mode device. No need to type in any URLs. I-Mode site in English: **http://imode.blink.com** I-Mode site in Japanese: **http://imode.blink.co.jp**
Cybird (**http://www.cybird.co.jp/**)	I-Mode services and charges.
DoCoMoZone (**http://www.bluga.com/docomo/**)	I-Mode, PHS, wireless, and much more. News, articles, and FAQs.
Eimode (**http://nuke.eimode.com/**)	News in English on i-Mode.
Fujitsu Japan (**http://jp.fujitsu.com/**)	Fujitsu Japan i-Mode site list (in Japanese).

continues

TABLE A-1 I-Mode Resources *(continued)*

RESOURCE	DESCRIPTION
Henry Minsky (**http://www.ai.mit.edu/people/hqm/imode/**)	I-Mode and Wireless Dev Notes.
I-Mode links (**http://imodelinks.com/desktop.html**)	I-Mode answers to FAQs.
imodeindia.com (**http://www.imodeindia.com/iappli/source/source.html**)	iAppli source code library.
Ithaki Metasearch (**http://www.ithaki.net/i/**)	I-Mode Metasearch engine. Searches for i-Mode sites in multiple search engines.
Japanese Web phones (**http://renfield.net/matrix.html**)	Matrix by service, network, and so on.
MobileBuddy.com (**http://www.mobilebuddy.com/**)	The MobileBuddy.com site is now compatible with i-Mode devices in addition to the previously supported WAP and Web browsers. Users only have to remember one URL (**http://www.mobilebuddy.com**) to access the site from any of these three types of browser. MobileBuddy.com provides access to Usenet news and POP3 mailboxes.
PersonalBuddy.com (**http://www.PersonalBuddy.com/**)	PersonalBuddy.com offers the same news and mail that have become familiar to MobileBuddy.com users, as well as mail forwarding and instant messaging via the Jabber.org messaging system.
Mobile Media Japan (**http://www.mobilemediajapan.com/**)	I-Mode FAQs.
Modezilla (**http://www.modezilla.com/index.php3**)	I-Mode, NTT DoCoMo, wireless developer news, and tools.
Network365 (**http://www.network365.com/**)	mZone mobile commerce server for WAP, i-Mode, and SMS.
NTT DoCoMo (**http://www.nttdocomo.co.jp/**)	I-Mode Java pages (in Japanese [**http://www.nttdocomo.co.jp/i/java/**]).
	I-Mode Java specs (in English [**http://www.nttdocomo.com/i/java/**]).
TechBuddha (**http://www.techbuddha.com/**)	Monetizing content: i-Mode content providers.

TABLE A-2 I-Mode Services

Resource	Service
Astel Mozio	About services (**http://web.mozio.net/ues.html**)
Au EZWeb	About services (**http://www.au.kddi.com/**)
DDI Pocket	About services (**http://www.ddipocket.co.jp/**)
J-Phone	About services (**http://www.j-sky.j-phone.com/**)
NTT DoCoMo	About services (**http://www.nttdocomo.com/top.shtml**)

TABLE A-3 I-Mode Phones and Other Compatible Devices

Resource	Device
I-Mode Eye (**http://www.japon.net/imode/**)	I-Mode devices (**http://www.japon.net/imode/imodehard.shtml**).
Japan Today (**http://www.japantoday.com/e/?content=home**)	Talking about my generation. Just when you thought you were on top of the technology driving mobiles and other flashy hi-tech accessories in Japan—here comes the third generation of goodies to get you wired. These are the products hot off the three top mobile phone companies' drawing boards . . . they're coming soon and yet to be priced.
Mobile Media Japan (**http://www.mobilemediajapan.com/**)	Top-selling handsets for NTT DoCoMo's i-Mode. Citing information from Telework Japan, Impress Keitai Watch has published a listing of the best-selling i-Mode-compatible handsets.
	List of available i-Mode handsets (**http://www.mobilemediajapan.com/hardware/imode-handsets/**).

continues

TABLE A-3 I-Mode Phones and Other Compatible Devices *(continued)*

RESOURCE	DEVICE
Nokia Japan (**http://www.nokia.co.jp/index.html**)	Nokia Japan phone products (**http://www.nokia.co.jp/products/**). Nokia debuts first i-Mode cellular phone. NTT DoCoMo Inc. starts selling its Digital MOVA NM502i HYPER cell phone supporting i-Mode, an Internet connection service. This is Nokia Mobile Communications KK's first i-Mode-supported phone. The Digital MOVA NM502i HYPER enables downloads of any image from Web sites and designates them as a standby screen. It is priced at 33,000 yen. (106.98 yen = US $1).
	DoCoMo Nokia NM502i mobile phone.
Nooper (**http://nooper.co.jp/showcase/**)	Visual Showcase—screen shots and photos of i-Mode phones.
NTT DoCoMo (**http://www.nttdocomo.com/top.shtml**)	F502i and D502i i-Mode phones.
Sun (**http://www.sun.com/**)	Fujitsu, Matsushita/Panasonic, Mitsubishi, and NEC demonstrate Java technology-based i-Mode wireless phones at JavaOneSM. Sun Microsystems and NTT DoCoMo deliver the first prototypes of NTT DoCoMo's Java technology-enabled i-Mode wireless phones and services. Powered by Sun's *K virtual machine* (KVM).
XML.com (**http://www.xml.com/**)	Getting into i-Mode. Introductory look at i-Mode.

NOTE

The first letters of the handset model indicates the manufacturer:

d = Mitsubishi — *f = Fujitsu*
n = NEC — *nm = Nokia*
p = Panasonic — *sh = Sharp*
so = Sony

MOBILE TECHNOLOGY RESOURCES

Table A-4 contains the following mobile technology resources:

- I-Mode/CHTML
- EZweb/WAP/HDML
- J-Sky/MML
- H"/PmailDX
- Palm/Windows CE
- JAVA
- Misc

TABLE A-4 Mobile Technology Resources

Technology Type	Resource
I-Mode/Compact HTML	DoCoMo Q&A on getting listed on iMenu (**http://discuss.mobilemediajapan.com/stories/storyReader$618**): NTT DoCoMo has posted a Q&A page (in Japanese) describing the procedure for getting a site listed as an official NTT DoCoMo iMenu destination. (WestCyber News Desk/English).
	Compact NetFront Microbrowser (**http://www.access-us-inc.com/product/develop/cn.html**): Background information about the HTML Microbrowser—the brain behind DoCoMo's i-Mode phones. Download a free evaluation version for Linux. (Access/English).
	English i-Mode FAQs (**http://imodelinks.com/desktop/faq.html**): Consumer-oriented general i-Mode FAQs (WestCyber/English).

continues

TABLE A-4 Mobile Technology Resources *(continued)*

Technology Type	Resource
	I-Mode mail (**http://www.nttdocomo.com/i/service/imail.html**): Description by NTT DoCoMo on how to use i-Mode e-mail. (NTT DoCoMo/English).
	I-Mode Q&A (**http://www.nttdocomo.com/i/qa.html**): Q&A about i-Mode, mail, mail settings, charges, site access, messages, and Internet access (NTT DoCoMo/English).
	HTML for information appliances (**http://www.webreview.com/1998/08_21/webauthors/08_21_98_1.shtml**): As technologies advance, we're going to see the Web popping up in more and more places. Companies are working with existing or slightly modified versions of HTML specifications to meet this demand. One such spec, Compact HTML, promises to bring the Internet from the big screen to small information appliances (Webreview.com/English).
	Tagset for Compact HTML (**http://www.webreview.com/1998/08_21/webauthors/08_21_98_2.shtml**): See which HTML 4.0 tags and attributes are Compact HTML-compliant (Webreview.com/English).
	Compact HTML for small information appliances (**http://www.w3.org/TR/1998/NOTE-compactHTML-19980209/**): The Compact HTML proposed here defines a subset of HTML for small information appliances such as smart phones, smart communicators, mobile PDAs, and and so on (W3C Submission/English).
	I-Mode FAQs (**http://www.sh.rim.or.jp/~kak/if/**): Frequently asked questions (and their answers) about i-Mode (in Japanese).

TECHNOLOGY TYPE	RESOURCE
	CHTML for i-Mode (**http://www.nttdocomo.co.jp/i/tag/index.html**): Tagset and structure of several versions of CHTML, explained by DoCoMo (NTT DoCoMo/Japanese). Also available in English, including a list of i-Mode picture symbols.
EZweb/WAP/HDML/WML	Phone.com HDML/WML Development Kit (**http://www.openwave.com/**): UP.SDK is a freely available software development kit that enables Web developers to quickly and easily create HDML and WML information services and applications. An HDML/WML simulator for the PC is included.
	Mobilizing the Web with HDML (**http://www.webreview.com/1998/10_09/webauthors/10_09_98_5.shtml**): Getting mobile no longer means being disconnected. See how the Handheld Devices Markup Language (HDML) promises to make the Web accessible to cellular phone users (Webreview.com/English).
	Proposal for a Handheld Device Markup Language (HDML) (**http://www.w3.org/Submission/**1997/5 /): HDML is a simple language used to define hypertext-like content and applications for hand-held devices with small displays (W3C Submission/English).
	WAP and WML: Delivering wireless content (**http://www.webreview.com/1999/10_22/strategists/10_22_99_4.shtml**): Introduction to WAP and WML (Wireless Markup Language).
	Handheld Device Markup Language FAQs (**http://www.w3.org/TR/NOTE-Submission-HDML-FAQ.html**): Frequently asked questions about HDML (W3C/English).

continues

TABLE A-4 Mobile Technology Resources *(continued)*

Technology Type	Resource
	HDML Homepages (**http://info.ezweb.ne.jp/custmize/**): How to make HDML pages (EZweb) and use the free creation software My Deck Editor (EZweb/Japanese).
J-Sky/MML	Mobile Markup Language (MML) for mobile network (**http://www.mobilemediajapan.com/resources/technology.html#chtml**): MML tag index (lower half of the page is in English).
	J-Sky-compatible HTML (**http://www.interq.or.jp/kansai/jackal/j/index.html**): The following HTML tags can be used as well to build J-Sky-compatible pages. Advantage: No need to learn MML, and the pages can be tested on desktop-based browsers like Netscape Communicator and Internet Explorer (Japanese).
	Homepage Guide for J-Sky-compatible HTML (**http://homepage1.nifty.com/opaken/hpguide/**): How to build J-Sky-compatible pages based on HTML, including background melodies and animation pictures (Japanese).
DDI Pocket (H"/PmailDX)	Open network content (**http://homepage1.nifty.com/opaken/hpguide/**): How to publish content on H"/PmailDX phones (DDI Pocket/Japanese).
	Open contents editor (**http://www.ddipocket.co.jp/download/opennet/easy_onc.html**): Free homepage builder open contents (page creation, bitmap conversion, and preview functionality) (DDI Pocket/Japanese/Windows 9x/NT/2000).
Palm/Windows CE	Becoming an AvantGo partner (**http://avantgo.com/support/developer/**): Make your content available to millions of handheld users by using simple HTML.

TECHNOLOGY TYPE	RESOURCE
	See also the *Developer Documentation, Tips and Tricks* and the *Style Guide for Handheld Devices* (Avantgo/English).
Misc	CSS Mobile Profile 1.0 (**http://www.w3.org/TR/2000/WD-css-mobile-20001013**): This specification defines a subset of the cascading style sheets level 2 specification tailored to the needs and constraints of mobile devices (W3C, English).
	XHTML Basic (**http://www.w3.org/TR/xhtml-basic/**): Subset of XHTML 1.1. It is designed for Web clients that do not support the full set of XHTML features; for example, Web clients such as mobile phones, PDAs, pagers, and settop boxes (W3C Working Draft, English).
	HTML 4.0 Guidelines for mobile access (**http://www.w3.org/TR/NOTE-html40-mobile/**): This document describes important guidelines to show content authors how to create HTML 4.0 contents to be acceptable to mobile devices as much as possible (W3C).
JAVA	The difference between iAppli and Sun's MIDP profile (**http://www.geocities.co.jp/SiliconValley-Cupertino/1621/**): Zev Blut has developed this site in the hopes of helping others who are getting started with developing iAppli programs for NTT DoCoMo's new mobile phones (Zev Blut/English).
	I-Mode Java specifications (**http://www.mobilemediajapan.com/newsdesk/imode-java.html**): NTT DoCoMo, in cooperation with Sun Microsystems (U.S.A.), Sun Microsystems K.K. (Japan), and handset manufacturers has developed a Java programming environment for i-Mode handsets (MMJ/English).

continues

TABLE A-4 Mobile Technology Resources *(continued)*

Technology Type	Resource
	I-Mode Java emulator (**http://uni-labo.com/products/iEmulator/**): Japanese emulator for the development of i-Mode Java-compatible applications (FUKUNO Taisuke/Japanese).
	Jblend (**http://www.jblend.com/en/**): JBlend has been adopted as the platform for the next generation cellular phone of J-PHONE (JBlend/English).
	JAVA technology goes wireless (**http://www.sun.com/sp/features/wireless.html**): How Sun and NTT DoCoMo are leading the mobile revolution (Sun Microsystems/English).
	Mobile information device profile for the J2ME platform (**http://jcp.org/jsr/detail/037.jsp**): This specification will define a profile that will extend and enhance the pJ2METM Connected, Limited Device Configuration (JSR-000030), enabling application development for mobile information appliances and voice communication devices (Sun Microsystems/English).
	Sun and NTT DoCoMo unveil prototypes of next-generation JAVA (**http://java.sun.com/pr/1999/06/pr990615-16.html**): Fujitsu, Matsushita/Panasonic, Mitsubishi, and NEC demonstrate Java technology-based i-Mode wireless phones at JavaOne (Sun Microsystems/English).
	The Java platform gains visionary allies (**http://java.sun.com/features/1999/03/wireless.html**): NTT DoCoMo and Symbian join the Java technology charge (Sun Microsystems/English).

WARNING

URLs are subject to change without notice.

APPENDIX B

JAPAN MOBILE INTERNET CASE STUDY: NTT DOCOMO I-MODE

The mobile Internet is becoming more and more prevalent in the Japanese market as each day passes, particularly with respect to NTT DoCoMo's i-Mode services. Because the wireline Internet penetration rates are relatively low in Japan compared with the United States, the Japanese are subscribing to the wireless Internet phenomenon.

I-Mode services are expanding very quickly in Japan and are potentially the key indicators regarding the future direction of the marketplace. Other wireless Internet providers are looking to i-Mode as a successful benchmark that should be emulated. This case study highlights the following market issues:

- Key mobile Internet services including NTT Mobile Communications Network Inc.'s (NTT DoCoMo) i-Mode services
- Dominant i-Mode services including linked gaming and downloading music from Internet to wireless phone
- Present and upcoming technologies involving the wireless Internet in Japan
- How NTT DoCoMo and its suppliers are expanding their domination of the mobile phone market to other markets[1]

AMAZING DOCOMO

The company's wireless Net phone service is all the rage in Japan—and just might conquer the world. However, a Japanese teenage girl doesn't leave home without a few things: her six-inch platform shoes, some touch-up toner for her hair color of the day, and her i-Mode phone.

Come again? Everyone in Japan knows i-Mode stands for a white-hot cell-phone service from NTT DoCoMo. The company name is a play on the Japanese word for anywhere, and the service lives up to that moniker by giving subscribers across Japan cheap and continuous wireless access to the Internet.

Europeans, too, can tap into the Net from their cell phones. Thousands of Americans get similar benefits using Palm VIII devices from 3Com Corp. However, all these systems must establish new dial-up connections each time a user wants to go onto the Internet. With i-Mode, users are always connected—as long as they can receive a signal, and their batteries are charged. Through this persistent link, subscribers get a full panoply of Web-based goodies: e-mail and chat, games, online horoscopes, calendars, and customized news bulletins. All told, i-Mode subscribers can navigate among 6,000 specially formatted Web sites.

In coming months, the rest of the world will be hearing a lot more about i-Mode—the service, the company that provides it, the stock, and the amazing individuals who helped create this phenomenon. Here are a few quick reference points: Since the company was partly spun off from NTT, Japan's former telephone monopoly, in 1992, DoCoMo has become

- Japan's hottest stock. We're talking a Berkshire Hathaway-level share price. In dollars, DoCoMo has soared from $13,000 a share to $57,000 since 2000.
- The world's most valuable cell-phone company, with a market cap of about $557 billion.

- The largest single-country cell-phone operator, with a total of 49.3 million Japanese subscribers.
- The most advanced wireless Net access service on the planet, period.

NTT's i-Mode is the only network in the world today that gives you (continuous) access to the Internet via a cell phone. More than five million Japanese have signed up, which means DoCoMo is now on its way to becoming the largest Internet service provider in Japan. DoCoMo has surprised a lot of people in the United States. You haven't seen anything like these numbers for any general wireless data service.

MASS MARKET

No wonder so many Japanese see DoCoMo as Japan's symbol for New Age innovation—the last, best hope to make Japan a contender in the global Internet derby. In the United States, companies, such as AT&T and Motorola Inc., are dickering over technical standards, the amount of bandwidth required for mobile data applications, and whether wireless Web surfers will prefer smart cell phones or palmtop computers. In Japan, DoCoMo has jumped far ahead, showing the world a mass market already, with millions of consumers who don't care that the technology isn't perfect.

In the future, DoCoMo plans to take on the likes of Vodafone AirTouch, AT&T, and British Telecom in markets around the world. Indeed, if the mobile Internet lives up to expectations, DoCoMo could evolve into the world's mightiest wireless giant. However, for that to happen, DoCoMo must prove that its new services and business models have staying power. Right now, i-Mode lets users perform all kinds of cool tricks. In Silicon Valley or Scandinavia, friends can't swap pictures of their pets on sleek, 90-g cell phones. Japanese do that every day.

However, pet tricks won't give DoCoMo a permanent edge over fierce global rivals. Already, European and U.S. cell-phone moguls are greedily eyeing Asia's potential market of 5.5 billion souls. Vodafone and British Telecom recently announced plans to launch a next-generation mobile service with Japan Telecom, a local carrier in which they both hold stakes.

Family Ties

Keiji Tachikawa, DoCoMo's charismatic chief executive, doesn't appear to be alarmed. He plans to fight the competition through a network of friendly alliances—a strategy that will likely include taking equity stakes in companies based mainly in Asia. Tachikawa is betting that hostile takeovers of the sort Vodafone has pursued will be a turn-off in Asia. Shunning those tactics, DoCoMo plans to woo Asian operators with funds and state-of-the-art technology.

Tachikawa has plenty of cash to power his expansion. DoCoMo's 2001 operating profits are expected to hit $7 billion, on revenues of $58 billion. Perhaps more important, DoCoMo represents a potent legacy—one that was once referred to as Japan Inc. Through the 1970s and 1980s, NTT and its family of powerful equipment suppliers (Fujitsu, NEC, and Hitachi among them) embodied some of Japan's most successful industrial policies.

This government-led industrial model looked like a liability during Japan's long recession. Today, newcomers like DoCoMo and its tightly knit content, equipment, and service suppliers can draw on those old family strengths. Giant NTT still owns 69 percent of DoCoMo. The market cap of both companies combined makes a hostile takeover unthinkable. At the same time, a stratospheric valuation could be currency for DoCoMo's own M&A designs down the road.

For Japanese makers of mobile equipment and handsets hard hit by recession, things haven't looked this promising in years. The local wireless market is expected to be worth $100 billion by 2003, versus $40 billion in 2001, according to the

Mobile Office Promotion Association. About 60 percent of the 2003 tally will be generated by wireless Internet services, excluding hardware.

Today's head start could hand Japan a leadership role when the mobile Net goes global. DoCoMo's innovators would reap licenses on standards they hewed. Their umbrella of Internet content suppliers and equipment makers would get a leg up in global markets. Japan has long lagged behind the United States in PC and Internet penetration, largely because of a lack of familiarity with the keyboard. However, personal electronics are another story. This is the country that gave the world the calculator, the Walkman, the pocket TV, the Game Boy, and the camcorder. Millions of Japanese grew up playing video and pocket computer games—the so-called push-button generation. Many are now migrating to Net-ready cellular handsets, often bypassing home computers altogether. They form a perfect testing ground for new Net appliances.

Headhunting

DoCoMo hasn't always been a hot property. In fact, it languished for years inside giant NTT. In 1992, when DoCoMo was partially spun off, few NTT staffers wanted to be assigned to the mobile carrier. At the time, the market was closed, subscription fees were costly, and the phones weighed as much as lunch boxes. What's more, a new technology known as personal communications services (handyphone in Japan) seemed poised to push aside other types of digital cellular systems.

In 1994, the postal ministry liberalized the cell-phone market, domestic competitors with names like DDI, IDO, and TU-KA blossomed, and prices for digital cell phones plummeted. DoCoMo's army of engineers (many inherited from NTT) fought the upstarts by developing the world's smallest cell phones, sharing the specifications with handset makers such as Fujitsu and Matsushita Communication. DoCoMo also blanketed the country with its branded retail shops and came up with a mascot called DoCoMo-chan.

All the while, as DoCoMo's conventional digital service was blossoming, DoCoMo's first president (Koji Ohboshi) was searching to extend the company's business beyond voice to data communications. That task fell to electrical engineer Keiichi Enoki. He did what few Japanese managers dare to do—headhunt. His first catch was Matsunaga, a senior executive at Recruit, and a publisher of job information, who had a track record of hatching successful businesses. With her help, he lured the third main team member, Takeshi Natsuno, a supercharged Internet entrepreneur who was running one of Japan's early online startups.

Mobile Model

After several months of brainstorming, the trio targeted wireless Internet access as the next big thing. DoCoMo engineers built a packet-switched network alongside their existing digital cellular network. With packet systems (as opposed to circuit-switched phone networks), each user doesn't need to receive an exclusive radio channel. That means many users can access the network at the same time. The packet model also reduces costs because charges are based on the volume of data sent and received.

Takeshi Natsuno devised the business model to make this system work. First, he dictated that i-Mode should serve as a portal site and lined up content providers that users could access directly from i-Mode's menu bar. Then he set up a billing method whereby DoCoMo would reap a commission for the services rendered by this first tier. Other content owners would be encouraged to code their Web pages for i-Mode as well. However, only those belonging to the licensed first tier could be accessed by the menu bar. This is a model for the mobile Internet that others now want to emulate.

Tachikawa, president and chief executive since 1998, insisted on a cheap pricing plan to guarantee the widespread adoption of i-Mode. Subscribers pay about four cents to send a 250-character message, and half that to receive a message of the same size. Tachikawa also insisted that the functions be as simple as possible.

The 62-year-old president arrived at DoCoMo with the right credentials. In addition to a Ph.D. in engineering from prestigious Tokyo University, he has an MBA from the Massachusetts Institute of Technology. An NTT executive for most of his career, he was responsible for drawing up several of NTT's long-term policy plans in the 1980s. He communicates with all DoCoMo personnel directly by e-mail—a practice still rare in Japan. He also manages to keep up with American baseball (he's a New York Mets fan) and football.

However, nothing Tachikawa ever said or did has rocked the roof like i-Mode. Since the service was launched in 2000, subscribers have been signing up at a rate of 762,000 a month. By the end of 2002, the number should hit 29 million. If the rate keeps accelerating, i-Mode could match America Online Inc.'s subscriber base of 43 million in 2003.

Toons and News

I-Mode has already become an addiction for millions of Japanese. The i stands for information—something young Japanese hunger for. Teens keep their handsets on round-the-clock and conduct rapid-fire exchanges of messages until their batteries give out. Students and young adults post passport-style pictures online, which can then be viewed or downloaded to their i-Mode handset screen. Bandai Co., the maker of Tamagotchi digital pets and animated films, operates a popular cartoon service, and some 800,000 subscribers pay about $1 a month for access to the site.

In its first year of operation, i-Mode is on track to post more than $100 million in revenues. That's small potatoes compared with sales of basic cell-phone services. However, Tachikawa estimates that a user base of seven million would generate $3.7 billion in fees from subscriptions, transmitted data, and a 9-percent commission DoCoMo charges for handling the billing for Web content providers.

The success of i-Mode sets the stage for DoCoMo's next big act—an evolutionary advance known around the planet as *third generation* (3G). In essence, it is a set of wireless protocols that will enable vastly higher communications speeds. Although the final standards have yet to be hashed out in America, Japan and Europe have committed to an approach called wideband CDMA, which will let users view streaming video and a host of other new Net applications at a blinding 2 Mbps by 2003—compared with just 9.6KB today.

In one sense, this migration may seem to neutralize DoCoMo's current advantage. After all, Europe and America will also jump to the new 3G protocols. The new services will be much flashier than anything you can do with today's i-mode phones. However, Tachikawa insists that DoCoMo will be able to carry its services and expertise along to the next generation, without exorbitant costs. All the excitement over DoCoMo today is just the beginning. Everyone will be on i-Mode in three years, according to Tachikawa.

To protect its lead, DoCoMo keeps up a brutal research pace. At a state-of-the-art research facility in the Yokosuka Research Park, southwest of Tokyo, about 700 engineers are testing transmission equipment, cellular phones, palmtops, and car navigation systems based on CDMA, which stands for *code division multiple access*. The technology was developed by the U.S. military and commercialized first by Qualcomm Corp. However, DoCoMo enhanced the basic platform with its own, homegrown mathematics. It got a big endorsement when Europe leaned in the same direction. Ultimately, Tachikawa hopes to convert other operators to DoCoMo's wideband CDMA system, inside and outside Japan.

By proselytizing for *wideband CDMA* (W-CDMA) around the world, Tachikawa hopes to extend i-Mode's reach without having to resort to takeovers. Recently, he negotiated a friendly deal to purchase a 19-percent stake in Hutchison Telecom, Hong Kong's largest cellular operator. Hutchison is expected soon to announce plans to adopt the i-Mode service and later on DoCoMo's 3G system. In hopes that others might follow suit, DoCoMo is also conducting 3G tests with operators in Malaysia, Singapore, and China, to be followed by possible investments.

Converts

However, Tachikawa is not ignoring the West. Recently, he struck a deal with Sun Microsystems to incorporate the company's Java program in i-Mode handsets coming out in the fall, as well as in the 3G devices scheduled to go on sale in 2002. Recently, Tachikawa and Microsoft President Steve Ballmer agreed to set up a joint venture, called Mobimagic, to develop wireless data services for the business market in Japan. If the new features and services take off in Japan, Tachikawa hopes to transplant them overseas. Recently, DoCoMo opened two U.S. subsidiaries in Silicon Valley—one for research and development and the other for promoting Japan's W-CDMA standard.

Some of DoCoMo's closest collaborators are already converts. Microsoft, for example, is doing most of its pioneering wireless data work with DoCoMo in Japan. Hopefully, that will migrate back to the United States as the wireless infrastructure improves.

As the mobile Internet takes off, DoCoMo will have to fight off competitors while entering new markets. Merger mania is already sweeping through the wireless world—much of it inspired by visions of the mobile Internet. Recently, there has been a flurry of alliances and mergers, such as MCI WorldCom Inc.'s buyout of Sprint Corporation. Britain's Vodafone AirTouch PLC is maneuvering to take over Germany's Mannesmann. The pressure from such massive consolidation could make competitors even more aggressive.

DoCoMo isn't immune at home. Rivals DDI Corp. and IDO Corp., which account for 27 percent of the cell-phone market, are merging their cellular operations to push a cellular standard that differs from DoCoMo's. The two are now introducing a packet network for a mobile Net service that will compete against i-Mode.

For now, however, DoCoMo has latched first and best onto the mobile Internet, a technology with far greater potential than the other portable-electronics markets Japan has conquered. Calculators and camcorders do not carry with them an entire set of complex, Internet-based services, complete with new business models and lush venture capital funding. All these and more come with the mobile Internet. Thanks to DoCoMo, Japan is out in front of the great land grab.

END NOTES

[1]Global Information Inc., 901 Farmington Ave., West Hartford, Connecticut, 06119, USA, 2001.

APPENDIX C

I-MODE INDEPENDENT RESOURCES: I-MODE EYE

As previously mentioned in earlier chapters of this book, every day, several hundred thousand Japanese switch on the world's only mobile phones with direct links to the Internet to say "Hello Kitty."

Japanese are queuing to buy i-Mode telephones that give the world's most technologically aware people the ability to download the childlike antics of Asia's most popular cartoon character, the pointy-eared, round-eyed, mouthless (and, above all, cute) Kitty Chan. Toymaker Bandai Co. and mobile phone operator NTT DoCoMo are gleeful. The Hello Kitty Web site, operated by Bandai, is one of the most popular sites on NTT DoCoMo's hugely successful i-Mode mobile Internet service, rapidly growing into Japan's principal platform for E-commerce.

KEEPING AN EYE ON I-MODE

As i-Mode attracts envious glances from Asian and European mobile carriers, it also promises to put Japan (long a laggard in Internet use) at the cutting edge of international cyber culture. A lot of people don't mind paying peanuts for such services; Bandai already earns huge revenues. The i-Mode service owes its popularity to a marriage of Japanese high-tech and pop culture, while enabling easy access to the Net. Japan's largest Internet provider just two years after its launch, the service is also one of the few business models for E-commerce to start generating profits so quickly. However, the release and recall of some i-Mode handsets has caused users and suppliers some anxiety (see sidebar, "Eye-to-Eye").

EYE-TO-EYE

RECALL

DoCoMo is asking testers to return the N2001 3G handsets due to problems connecting and keeping a connection to the network. One thousand four hundred (1,400) terminals are affected by this action, and DoCoMo seems to take the viewpoint that it is to be expected on a testing period. The camera-able P2101V also has some less serious bugs, and 1,200 terminals are still to be delivered to the testers.

CRIME AND PUNISHMENT

Bugs forcing a recall are the crime, and the punishment is on the left: In previous recalls, the handset involved was pulled out of shops; this time DoCoMo is also taking it out of its printed materials. This is for a temporary recall. Those in publishing will clearly understand the increased zeal demonstrated by the carrier.

I-MODE – MINITEL? L-MODE

Unless the courts stop it, L-Mode is being released very soon. The wired siblings of NTT DoCoMo, NTT-East and NTT-West have not been blind to the i-Mode success and L-Mode is their answer.

The name is not the only similarity, it is technically based on IP, cHTML and—at least initially—small monochrome LCD screens. With an official menu lined up with a who's who of the Japanese mobile Web, differences abound as the devices are targeted towards the home, seniors and housewives, and the contents. Conservative and boring, the NHK of the information age would be a somewhat unfair label, but the tons of documents that East and West are making official sites fill-in attest that these are more Telecom than Internet companies. Nevertheless, the risk of another dominant carrier using a service to further dominate the market is what is pushing secondary carriers, such as KDDI, to threaten with a lawsuit. As with the current NTT law, East and West are officially barred from Internet (IP) business.

3G IN THE STREET, IS IT WORKING?

Handsets are finally arriving in the hands of a few lucky testers in the greater Tokyo area. So? Bugs!

Complex i-Mode pages and not-so-complex Java i-applis are not working. This occurs when signals are reaching the phones, which is not always the case. Real terminals for independent testing are in extremely short supply while bugs are being worked out. It's a good thing that it is a test period, as the former MPT is making noises by not enabling 3G data transmissions until quality improves. Time will tell.

REAL 3G HANDSETS

Recently, the world's first 3G handsets found their way into the hands of 4,000 testers. However, you do not have to wait that long to see the new phones. They look nothing like the mock-ups that have circulated in trade fairs and selected locations.

- **P2101V by Panasonic** Outwardly similar to the P503is, it sports a camera that, besides taking stills, enables it to function as a TV phone with other P2101V handsets.
- **N2001 by NEC** The standard phone, with an improved color screen and (like the P2101V) no external antenna.
- **P2401 by Panasonic** A PCMCI card designed for data transmission up to 384 Kpbs downstream and 64K upstream.

Users will be able to talk and access contents/e-mail simultaneously. While completely compatible with current i-Mode contents, i-applis can grow in size from a max of 10K to 50K. Regular handsets include a smart card similar to the one found in GSM phones and should ditch the current serial connection for a mini-USB port.

continues

continued

NEW J-PHONE CAMPAIGN: NAKATA

Hide Nakata, captain of the National Soccer Team, star player of the Roma F.C., and all-around personality in Japan is the new face of J-Phone. The brand new TV and regular media campaign centers on a "Be Yourself" theme, with Nakata giving a monologue on the need to choose (things), not just to be different, but because they fit ones lifestyle (look beyond DoCoMo).

Nakata (nakata.net), an independent and stylish image, most notably also representing CANON, is gradually replacing actress Fujiwara Norika. NTT DoCoMo has, since the start of i-Mode, used young actress Hirosue's pretty face and actor Tamura's somewhat quizzical appearances while KDDI is focusing on making AU a fashion label.

PANASONIC AND NEC: THE FIRST 3G FOMA HANDSETS

FOMA testing includes lots of information. However, of unique value was that the code of the handsets to be readily available. P and N are the starting letters. That means that the two closest associates of NTT DoCoMo will be supplying the first 3G Handsets in the world. No mock-ups, no hand-made units, but real, mass-produced terminals.

P503IS LAUNCHED

During the ending weekend of the Japanese Golden Week vacation, the clamshell Panasonic P503i arrived in the shops. It's a great looking machine that is hoping to ride the success of the best selling P209is, but with a relatively disappointing 256 color screen. It has the following features:

- 256 color screen
- 16 sound tones
- Monochrome LCD display private window
- IR port Ir-kiss
- I-appli, ssl, PlayStation-ready
- 98 grams

- Autonomy: 145 min/440 hrs
- Optional, detachable mini keyboard
- Bilingual (as all 503i)
- Street price: around 31.000Y (below Sony and NEC)

The p503is arrived weeks behind original schedule. The street points to increased quality control scrutiny from DoCoMo in order not to repeat the P503i fiasco.

VODAFONE ACQUIRES BT STAKES IN J-PHONE

The news that BT was selling all its assets at Japan Telecom and wireless subsidiary J-Phone have been confirmed. BT's debt is forcing a *sayonara* to the world's leading mobile market and a very much increased role for Vodafone.

DOCOMO 3G: DELAYED UNTIL 2002

DoCoMo delayed its 3G rollout to 2002 due to network issues. Japanese new sites are not updated yet.

This comes a few weeks after J-Phone announced their roll out had also changed to 2002 due to cost and international compatibility issues. So now it looks like the Isle of Man will be the first to have 3G.

UPDATE

A few technical issues will delay the commercial launch, including network reliability, handset availability, and general quality control. Clearer are the market reasons: little demand for the current 64K-PHS video and music offerings, and great demand for the 9.6K i-appli color handsets that were introduced in late January 2001. A big launch would likely reduce the 503i series sales, although it is not clear consumers would flock to the new products.

SONY JAVA PHONE: BUG!

Although DoCoMo is keeping quiet about this one, the new Sony SO503i is getting quite a number of consumer complaints centered on problems with the ring tone and sudden disconnections while talking

continues

continued

as well as marks left on the TFT screen. Over 900 claims have been made just in the southern region of Kyushu. Meanwhile, DoCoMo is not issuing any warnings or recalls and is instead offering exchanges or repairs on an individual basis.

This problem comes after serious faults with the Panasonic P503i of the same Java-capable series and the recall of several handsets manufactured in 2000. It is expected that DoCoMo will increase its quality control efforts on the handset manufacturers.

THE FUTURE OF JAPANESE E-COMMERCE

The i-Mode keeps users continually linked to the Internet, enabling them to exchange e-mail, swap pictures, call up restaurant guides, and navigate among 5,800 specially formatted Web sites without having to dial-up each time. The statistics leave no doubt about its popularity.

Every day, a total of 14 million e-mail messages are exchanged by its 6.4 million subscribers. They use the Net an average of 12 times a day, and more than 82 percent pay for some sort of content.

DoCoMo plans to incorporate i-Mode functions into all of its handsets and expects i-Mode users to reach 40 million by the end of 2002, matching America Online Inc.'s current subscriber base of 41 million.

DoCoMo also earns money from E-commerce activity on its platform. It takes a 9-percent commission on billing for its Web site operators in addition to subscription fees and data transmission fees from subscribers.

The i-Mode service brings DoCoMo an additional 2,000 yen in revenues a month per subscriber. That's a little more than kitty litter for DoCoMo (the world's second-biggest mobile phone operator with 5.8 trillion yen [$55.73 billion] in annual revenues), but the unique mobile Internet service helped it fend off rivals and dominate Japan's booming mobile phone market.

I-Mode Spawns Start-Ups

I-Mode's success has galvanized content providers for mobile communications tools while spawning a number of agile start-ups, including browser developers. Many are looking to expand their business abroad as Asian and European mobile phone carriers eye their own wireless data services modeled after i-Mode.

In Hong Kong, Hutchison Whampoa Ltd.'s mobile phone arm, in which DoCoMo owns a 19-percent stake, is expected soon to announce plans to adopt the i-Mode service. Inquiries are flooding in from carriers and mobile phone makers in South Korea, Taiwan, and Hong Kong, asking to provide them content for their planned mobile Internet services.

For example, the Bandai i-Mode cartoon site that includes Kitty (the pink, mouthless cat with a huge bow on her ear), now generates nearly $33 million in annual revenues from its 1,172,000 members, who pay about $1 a month. Content producer Cybird Co., which offers surfers (in the ocean) up-to-date wave information from around the world, indicates it is in talks with several large Asian carriers on equity tie-ups. It recently sold an equity stake to Intel Corp. for an undisclosed sum, as the U.S. chip titan seeks to strengthen its ties with cell-phone companies.

Silicon Valley venture capitalists eager for investment opportunities are also chasing after Takateru Imaizumi, a 29-year-old programmer who designed a search-engine for i-Mode. I-Mode is getting smarter.

DoCoMo plans to incorporate Sun Microsystems Inc.'s Java programming into i-Mode handsets coming out in 2002. That will enable animated figures to move more smoothly or will enable automatic updating of daily news, stock prices, and other information once it has been downloaded. You'll be able to do many things, like having Kitty Chan moving back and forth across the display and fetching the latest news for you every day.

APPENDIX D

CREATING I-MODE-COMPATIBLE WEB SITES

This appendix provides a very basic explanation of how to create i-Mode-compatible Web sites. The only requirement for making a Web site viewable by i-Mode terminals is that it be created using i-Mode-compatible HTML (see an outline of i-Mode-compatible HTML at: **http://www.nttdocomo.com/i/tag/imodetag.html**). Once the Web site is made, you need only upload it to an Internet Web server, just as you would with normal Web sites. The following are i-Mode Web site specifications and resources:

- I-Mode Web sites use subsets of HTML 2.0, 3.2, and 4.0.
- S-JIS character encoding must be used. Images should be in the GIF format.
- Java and other scripting languages are not supported.
- Half-width kana characters can be used.
- A list of i-Mode-compatible HTML tags (**http://www.nttdocomo.com/i/tag/lineup.html**) are also shown in Tables D-1, D-2, D-3, and D-4.
- A list of picture symbols (**http://www.nttdocomo.com/i/tag/emoji/index.html**) are also shown in Table D-5.

- The official i-Mode menu contents criteria (**http://www.nttdocomo.com/i/tag/criteria.html**).
- An application for a new information provider (**http://www.nttdocomo.com/i/tag/newip.html**).

NOTE

NTT DoCoMo does not offer Web site development services.

TABLE D-1 I-Mode-Compatible HTML 1.0 Tag/Screen Image Correspondence Table

<!---->	<A>	ACCESSKEY ATTRIBUTE DIRECTKEY FUNCTION
<BASE>	<BLOCKQUOTE>	<BODY>
 	<CENTER>	<DIR>
<DL, DT, DD>	<DIV>	<FORM>
<HEAD>	<H>	<HR>
<HTML>	<IMG>	<INPUT>
<LI>	<MENU>	<OL>
<OPTION>	<P>	<PLAINTEXT>
<PRE>	<SELECT>	<TEXTAREA>
<TITLE>	<UL>	

TABLE D-2 I-Mode-Compatible HTML 2.0 Tag/Screen Image Correspondence Table

<!---->	<A>	ACCESSKEY ATTRIBUTE DIRECTKEY FUNCTION
<BLINK>	istyle attribute	<MARQUEE>
<META>	<SELECT>	<OL>
<LI>	<BODY>	<FONT>
multiple attributes	cti attributes	

TABLE D-3 Other Functions Supported by Handset 502i Series

FUNCTION	DESCRIPTION
Animation GIF	Color palette
Automatic image size reduction	Line feed code
Handling of space	Special characters

TABLE D-4 I-Mode-Compatible HTML 3.0 Tag/Screen Image Correspondence Table

<!---->	<A>	ACCESSKEY ATTRIBUTE DIRECTKEY FUNCTION
<OBJECT>	utn	subject body
telbook kana e-mail		

TABLE D-5 GIF Support

GIF TYPE	I-MODE-COMPATIBLE HTML VERSION		
	3.0	2.0	1.0
Non-interlaced GIF	O	O	O
Interlaced GIF	O	O	X
Transparent GIF	O	Δ	X
Animation GIF	O	O	X
JPEG	□	–	–

Δ Supported only by color-display models.
X Displayed as a non-interlaced GIF.
□ Only on models supporting JPEG.
– Unsupported.
O Supported.

NOTE

*The correspondence tables provided are to show how individual tags appear on i-Mode terminals. Illustrations of screen images are for example only. Actual images may differ depending on the terminal manufacturer. Symbols 0 to 9, #, and * can be used as accesskeys for the directkey functions discussed in the correspondence table. HTML tag attributes can either be in half-space characters, full-space characters, or a combination of both.*

APPENDIX E

INTERNET, M-COMMERCE, I-MODE IN JAPAN

Like many other areas, several different agencies, ministries, and others collect statistics on Internet usage in Japan. Results differ a lot. Here are some current estimates:

- Japan's Internet users (from data at the end of August, 2001): 49 million, an 81.5 percent increase compared to one year ago.
- NTT DoCoMo's i-Mode (i-Mode, mobile Internet access): 27,264,000 i-Mode subscribers (September 14, 2001, increasing 60,000 to 70,000 per day).
- NTT DoCoMo's i-Mode: 73,296,000 subscribers (August 31, 2001, cell: 67,429,000 + PHS: 6,978,000).

M-COMMERCE

According to Japan's NTT DoCoMo, m-commerce is expected to generate $322 billion worldwide by 2006. This statistic, along with aggressive wireless Web development, demonstrates that something is in the air. Still, many skeptics of the wireless Web feel that the coming 3G cellular technology will not be the answer.

Some are waiting for wireless standards to be firmly established before building out 3G networks. Predicted data rates of up to 2 Mbps are already being scaled back, and carriers are unlikely to lower their service fees after investing billions in licensing 3G frequencies. Limited radio spectrum and poor indoor coverage are also impeding enthusiasm, as well as the consumer's growing disenchantment.

One incentive for m-commerce development in Japan is, of course, the U.S. FCC's E911 mandate, which will require cell phone companies to deploy wireless device locating technologies (such as GPS) within 100 meters of use. This has prompted location-based technology companies to develop enhanced services beyond the emergency variety, and WAP providers are now forging partnerships with carriers to bring ubiquitous services to the consumer. Content providers will, in turn, have greater reach for widespread distribution, giving local businesses a little boost on slower days. It will also drive more opt-in traffic and increased minutes as consumers look forward to location-based deals on sports or theater tickets, time-sensitive digital coupons, and so on.

To start with, there is a difference between m-commerce (B2C) and m-business (B2B), and the industry is somewhat divided on the issue of where to focus their energies. Some feel that the initial consumer fascination (and revenue) with mobile devices is already on the wane, and the vast majority of commerce will be in B2B rather than B2C applications. NTT DoCoMo indicates that enterprises must enable complete mobile capability for customer and employee interaction, or else risk becoming non-competitive. This is just a case of making sure the horse is put in front of the cart. NTT DoCoMo has also indicated that the novelty of mobile business solutions will not be what drives adoption—realistic, usable applications that present enough value to the enterprise will drive market acceptance. In other words, once an enterprise has figured out how to help themselves with m-business, it can then develop that experience further to help consumers with m-commerce. Yes, it is another channel to be managed, but also another channel by which to reach the consumer.

APPENDIX F

COMPLETE TAGSET OF I-MODE-COMPATIBLE HTML TAGS IN JAPAN

The complete tagset of i-Mode-compatible cHTML tags is shown in Table F-1:

TABLE F-1 Complete Tagset of I-Mode-Compatible cHTML Tags

cHTML Tags	Attribute	Function
!-- [Version 1.0]	--	Designates a comment.
&XXX; [Version 1.0]	--	Designates a value. For example, &, >, ", , and �
A [Version 1.0]	name=	Designates the marker name within an HTML file.
	href="URL"	Designates a link to a Web site (http), e-mail address (mailto), or phone number (tel).
	accesskey="char"	Directkey function. Designates handset buttons for menu selection. Allowable values for char: "0" to "9".
.BASE [Version 1.0]	href="URL"	Designates the base URL for the relative paths used in an HTML file.

continues

TABLE F-1 Complete Tagset *(continued)*

cHTML Tags	Attribute	Function
BLOCKQUOTE [Version 1.0]	--	Indents a block of text.
BODY [Version 1.0]	--	Designates content to be displayed as a page.
	bgcolor=	Supported by color-capable models only (cHTML Version 2.0).
	text=	Supported by color-capable models only (cHTML Version 2.0).
	link=	Supported by color-capable models only (cHTML Version 2.0).
BR [Version 1.0]	clear=all | left | right	Designates the way a character string wraps around an inline image by deciding where line feeding takes place. Depending on the attribute, it also cancels the wraparound function.
CENTER [Version 1.0]	--	Centers character strings, images, and tables.
DD [Version 1.0]	--	Creates a definition list (see DL).
DIR [Version 1.0]	--	Creates a list of menus or directories. The list is nested between <DIR> and </DIR>, and each list item must be preceded with <LI> just as in the list created by the UL tag.
DIV [Version 1.0]	align=left | center | right	<DIV ALIGN="left"> aligns the tagged text block to the left. <DIV ALIGN="center"> centers the tagged text block. <DIV ALIGN="right"> aligns the tagged text block at right.

cHTML Tags	Attribute	Function
DL [Version 1.0]	--	Creates a definition list. The entire list must be between <DL> and </DL>. Tags, such as <DT> and <DD>, are used between <DL> and </DL> to define the list in greater detail.
DT [Version 1.0]	--	Designates the list heading and aligns the character string to the left.
FORM [Version 1.0]	action=	Encloses an area to be shown as a data input form. The tag should be followed by the URL or e-mail address (mailto) the input form will be sent to.
	method=get | post	Designates the method by which data is sent to the server, to either post or get.
HEAD [Version 1.0]	--	Designates the information that is used as the page title and/or by the server. The HEAD tag follows the HTML tag.
Hn [Version 1.0]	--	Designates the size of the header.
Bb	align=left | center | right	Designates the alignment of the header.
HR [Version 1.0]	--	Designates the settings of the horizontal dividing line.
	align=left | center | right	Designates the alignment of the horizontal line.
	size=	Sets the thickness of the horizontal line.
	width=	Determines the length of the horizontal line.
	noshade	Gives the horizontal line a two-dimensional appearance. Please note that the noshade attribute is currently not supported.

continues

TABLE F-1 Complete Tagset *(continued)*

cHTML Tags	Attribute	Function
HTML [Version 1.0]	--	Indicates that a character string placed between <HTML> and </HTML> is an HTML file.
IMG [Version 1.0]	src=	Designates an image file (obligatory attribute).
	align=top | middle | bottom	Defines the way the image and character string are laid out and how the character string wraps around the image.
	align=left | right	Defines the way the image and character string are laid out and how the character string wraps around the image.
	width=	Sets the image width.
	height=	Sets the image height.
	hspace=	Sets the blank space to the left of the image on the screen.
	vspace=	Sets the blank space between the image and the preceding line.
	alt=	Designates a text string that can be shown as an alternative to the image.
	border=	Defines the image frame. Please note that the border attribute is currently not supported.
INPUT [Version 1.0]	type=text	Displays a textbox.
	name=	Designates the name of the field employed to pass the data, obtained using the INPUT tag, to a CGI script and others.
	size=	Designates the width of the textbox by number of characters.
	maxlength=	Limits the number of characters that can be input to the textbox.

cHTML Tags	Attribute	Function
	accesskey="char"	Directkey function.
	value=	Designates the initial value of the data.
	istyle=	Sets the default character input mode. 1=full-space kana, 2=half-space kana, 3=alphabetic, 4=numeric (supported in HTML Version 2.0).
	type=password	Displays a password-input textbox.
	name=	Designates the name of the field employed to pass the data, obtained using the INPUT tag, to a CGI script and others.
	size=	Designates the width of the password-input textbox by number of characters.
	maxlength=	Limits the number of characters that can be input to the password-input textbox.
	accesskey="char"	Directkey function.
	value=	Designates the initial value of the data.
	type=checkbox	Displays a checkbox.
	name=	Designates the name of the field that is used to pass the data, obtained using the INPUT tag, to a CGI script and others.
	value=	Designates the initial value of the data.
	accesskey="char"	Directkey function.
	checked	Makes a selected checkbox the default.

continues

TABLE F-1 Complete Tagset *(continued)*

cHTML Tags	Attribute	Function
	type=radio	Displays a checkbox.
	name=	Designates the name of the field that is used to pass the data, obtained using the INPUT tag, to a CGI script and others.
	value=	Designates the initial value of the data.
	accesskey="char"	Directkey function.
	checked	Makes a selected radio the default.
	type=hidden	Transmits the data designated by the value attribute immediately following "field name=."
	name=	Designates the name of the field employed to pass the data, obtained using the INPUT tag, to a CGI script and others.
	value=	Designates the initial value of the data.
	type=submit	Displays a submit button.
	name=	Designates the name of the field employed to pass the data, obtained using the INPUT tag, to a CGI script and others.
	accesskey="char"	Directkey function.
	value=	Designates the name of the button.
	type=reset	Displays a submit button.
	name=	Designates the name of the field employed to pass the data, obtained using the INPUT tag, to a CGI script and others.
	accesskey="char"	Directkey function.
	value=	Designates the name of the button.

cHTML Tags	Attribute	Function
LI [Version 1.0]	--	Changes the symbol that precedes a sentence.
	type=1 | A | a	1: Labels with numbers.
		a: Labels with lowercase letters.
		A: Labels with uppercase letters (supported in cHTML Version 2.0).
	value=	Defines the start attribute for labeling (supported in cHTML Version 2.0).
MENU [Version 1.0]	--	Creates a menu list.
OL [Version 1.0]	--	Creates a numbered list.
	type=1 | A | a	1: Labels with numbers.
		a: Labels with lowercase letters.
		A: Labels with uppercase letters (supported in cHTML Version 2.0).
	value=	Defines the start attribute for labeling (supported in cHTML Version 2.0).
OPTION [Version 1.0]	selected	Designates the selected (initial value) in a select list.
	value=	--
P [Version 1.0]	--	Creates a text block.
	align=left | center | right	Designates the alignment of the text block.
PLAINTEXT [Version 1.0]	--	Displays a text file exactly as entered. When the text file has more characters than can fit in the screen width, it will be broken into multiple lines automatically.
PRE [Version 1.0]	--	Displays a source file exactly as entered, including line feeds and blank spaces. When the

continues

TABLE F-1 Complete Tagset *(continued)*

cHTML Tags	Attribute	Function
		source file has more characters than can fit in the screen width, it will be broken into multiple lines automatically.
SELECT [Version 1.0]	name=	Designates the name of the list for passing selected items.
	size=	Designates the number of lines for the list.
	multiple	Enables multiple selections (supported in cHTML Version 2.0).
TEXTAREA [Version 1.0]	name=	Designates the name of the field employed to pass the data, obtained using the TEXTAREA tag, to a CGI script and others.
	rows=	Designates the height of the input box field.
	cols=	Designates the width of the input box field.
	istyle=	Sets the default character input mode.
		1=full-space kana,
		2=half-space kana,
		3=alphabetic,
		4=numeric (supported in cHTML Version 2.0).
TITLE [Version 1.0]	--	Designates the page title.
UL [Version 1.0]	--	Creates a bullet point list.
META [Version 1.0]	--	Supports only the definition of character codes. (I-Mode uses shift-JIS and Western ISO-8859-1 characters only.)

cHTML Tags	Attribute	Function
	http-equiv=	http-equiv=Content-Type (cHTML Version 2.0). Note that i-Mode terminals do not support the http-equiv="refresh" attribute.
	content=	content=text/html (cHTML Version 2.0).
	charset=	charset=shift-JIS (cHTML Version 2.0).
	name=	name="CHTML" content="yes" name="description" content="description of the site's content".
BLINK [Version 1.0]	--	Blinks a tagged character string at intervals. Some i-Mode terminals stop blinking the text after a predetermined period of time has elapsed to conserve power. BLINK will not be recognized by models that do not support the tag (supported in cHTML Version 2.0).
MARQUEE [Version 2.0]	behavior=scroll | slide | alternate	Runs a character string with a maximum length of 64 bytes on the screen.
	direction=left | right	--
	loop=	Up to 16 times.
	height=	Fixed and cannot be altered by users.
	width=	Fixed and cannot be altered by users.
	scrollamount=	Cannot be altered by users.
	scrolldelay=	Cannot be altered by users.
FONT [Version 2.0]	color=	Supported by color capable models only (cHTML Version 2.0).

APPENDIX G

COMPACT HTML TAG LIST

The list of cHTML tags is shown in Table G-1.

TABLE G-1 cHTML Tags

No	Elements	Attributes	HTML	CH	Comments
1	!-	-	2.0	CH	--
2	!DOCTYPE	-	2.0	CH	--
3	&xxx;	-	2.0	CH	--&, ©, >, <, ", ®, --�?'
4	A	name= href="URL" rel= rev= title= urn=(deleted from HTML 3.2) methods=(deleted from HTML 3.2)	2.0	CH CH - - - - -	--
5	ABBR	-	4.0	-	--
6	ACRONYM	-	4.0	-	--
7	ADDRESS	-	2.0	-	--Only one font.
8	APPLET	-	3.2	-	--(Deprecated element in HTML 4.0.)
9	AREA	shape= coords= href="URL" alt= nohref	3.2	-	--

No	Elements	Attributes	HTML	CH	Comments
10	B	-	2.0	-	--Only one font.
11	BASE	href="URL"	2.0	CH	--
12	BASEFONT	size=	3.2	-	--Only one font.
					--(Deprecated element in HTML 4.0.)
13	BDO	-	4.0	-	--
14	BIG	-	3.2	-	--Only one font.
15	BLOCKQUOTE	-	3.2	CH	--
16	BODY	-	2.0	CH - -	--Nonwhite colors are drawn as black.
		bgcolor=	3.2	- - - -	
		background=	3.2		
		text=	3.2		
		link=	3.2		
		vlink=	3.2		
		alink=	3.2		
17	BR	-	2.0	CH CH	--
		clear=all/left/right	3.2		
18	BUTTON	-	4.0	-	--
19	CAPTION	-	3.2	-	--

continues

TABLE G-1 cHTML Tags (*continued*)

No	Elements	Attributes	HTML	CH	Comments
20	CENTER	-	3.2	CH	--(Deprecated element in HTML 4.0.)
21	CITE	-	2.0	-	--Only one font.
22	CODE	-	2.0	-	--Only one font.
23	COL	-	4.0	-	--
24	COLGROUP	-	4.0	-	--
25	DD	-	2.0	CH	--
26	DEL	-	4.0	-	--
27	DFN	-	3.2	-	--
28	DIR	- compact	2.0	CH -	--(Deprecated element in HTML 4.0.)
29	DIV	- align=left/center/right	3.2	CH CH	--
30	DL	- compact	2.0	CH -	--
31	DT	-	2.0	CH	--

No	Elements	Attributes	HTML	CH	Comments
32	EM	-	2.0	-	--Only one font.
33	FIELDSET	-	4.0	-	--
34	FONT	size=n size=+n/–n color=	3.2	- - -	--Only one font. --(Deprecated element in HTML 4.0.)
35	FORM	action= method=get/post enctype=	2.0	CH CH CH	--
36	FRAME	-	4.0	-	--(Frameset DTD)
37	FRAMESET	-	4.0	-	--(Frameset DTD)
38	HEAD	-	2.0	CH	--
39	Hn	- align=left/center/right	2.0 3.2	CH CH	--
40	HR	- align=left/center/right size= width= noshade	2.0 3.2 3.2 3.2 3.2	CH CH CH CH CH	--

continued

TABLE G-1 cHTML Tags *(continued)*

No	Elements	Attributes	HTML	CH	Comments
41	HTML	-	2.0	CH CH	--version="C-HTML 1.0."
		version=	3.2		
42	I	-	2.0	-	--Only one font.
43	IFRAME	-	4.0	-	--(Frameset DTD.)
44	IMG	src=	2.0	CH CH	--Large images compressed automatically.
		align=top/middle/bottom align=left/right	2.0	CH CH	
		width=	3.2	CH CH	
		height=	3.2	CH CH	
		hspace=	3.2	CH	
		vspace=	3.2	-	
		alt=	3.2	-	
		border=	2.0		
		usemap=	3.2		
		ismap=	3.2		
			2.0		
45	INPUT	type=text	2.0	CH CH	--Max character buffer 512 bytes.
		name=		CH CH	
		size=		CH	
		maxlength=			
		value=			

No	Elements	Attributes	HTML	CH	Comments
		type=password	2.0	CH CH	--
		name=		CH CH	
		size=		CH	
		maxlength=			
		value=			
		type=checkbox	2.0	CH CH	--
		name=		CH CH	
		value=			
		checked			
		type=radio	2.0	CH CH	--
		name=		CH CH	
		value=			
		checked			
		type=hidden	2.0	CH CH	--
		name=		CH	
		value=			
		type=image	2.0	-	--
		name=	2.0		
		src= align=top/middle/bottom/left/right	2.0		
			3.2		
		type=submit	2.0	CH CH	--
		name=		CH	
		value=			

continues

TABLE G-1 cHTML Tags *(continued)*

No	Elements	Attributes	HTML	CH	Comments
		type=reset name= value=	2.0	CH CH CH	--
		type=file name= value=	3.2	-	--
46	INS	-	4.0	-	--
47	ISINDEX	- prompt=	2.0 3.2	- -	--(Deprecated element in HTML 4.0.)
48	KBD	-	2.0	-	--Only one font.
49	LABEL	-	4.0	-	--
50	LEGEND	-	4.0	-	--
51	LI	- type=1/A/a/I/i type=circle/disc/square value=	2.0 3.2 3.2 3.2	CH - - -	--

No	ELEMENTS	ATTRIBUTES	HTML	CH	COMMENTS
52	LINK	href="URL" rel= rev= urn= methods= title= id=	2.0	-	--
53	LISTING	-	2.0	-	--Only one font. --(Obsolete element in HTML 4.0.)
54	MAP	name=	3.2	-	--
55	MENU	- compact	2.0	CH -	--(Deprecated element in HTML 4.0.)
56	META	name= http-equiv= content=	2.0	CH	--http-equiv="refresh" only.
57	NEXTID	n=	2.0	-	--Deleted from HTML 3.2.
58	NOFRAMES	-	4.0	-	--(Frameset DTD.)
59	NOSCRIPT	-	4.0	-	--
60	OBJECT	-	4.0	-	--

continues

TABLE G-1 cHTML Tags (*continued*)

No	Elements	Attributes	HTML	CH	Comments
61	OL	-	2.0	CH	--
		type=1/A/a/I/i	3.2	-	
		start=	3.2	-	
		compact	2.0	-	
62	OPTGROUP	-	4.0	-	--
63	OPTION	-	2.0	CH CH	--
		selected		-	
		value=			
64	P	-	2.0	CH CH	--
		align=left/center/right	3.2		
65	PARAM	-	4.0	-	--
66	PLAINTEXT	-	2.0	CH	--(Obsolete element in HTML 4.0.)
67	PRE	-	2.0	CH	--
		width=	3.2	-	
68	Q	-	4.0	-	--
69	S	-	2.0	-	--(Deprecated element in HTML 4.0.)
70	SAMP	-	2.0	-	--Only one font.

No	Elements	Attributes	HTML	CH	Comments
71	SCRIPT	-	3.2	-	--
72	SELECT	name= size= multiple	2.0	CH CH CH	-Max character buffer 4 KB.
73	SMALL	-	3.2	-	--Only one font.
74	SPAN	-	4.0	-	--
75	STRIKE	-	2.0	-	--(Deprecated element in HTML 4.0.)
76	STRONG	-	2.0	-	--Only one font.
77	STYLE	-	2.0	-	--
78	SUB	-	3.2	-	--
79	SUP	-	3.2	-	--
80	TABLE	- align=left/center/right etc. border= width= cellspacing= cellpadding=	3.2	-	--
81	TBODY	-	4.0	-	--

continues

TABLE G-1 cHTML Tags (*continued*)

No	Elements	Attributes	HTML	CH	Comments
82	TD	- align=left/center/right valign=top/middle/bottom/baseline rowspan= colspan= width= height= nowrap	3.2	-	--
83	TEXTAREA	name= rows= cols=	2.0	CH CH CH	--Max character buffer 512 bytes.
84	TFOOT	-	4.0	-	--
85	TH	- align=left/center/right valign=top/middle/bottom/baseline rowspan= colspan= width= height= nowrap	3.2	-	--
86	THEAD	-	4.0	-	--
87	TITLE	-	2.0	CH	--

No	Elements	Attributes	HTML	CH	Comments
88	TR	-	3.2	-	--
		align=left/center/right valign=top/middle/bottom/baseline			
89	TT	-	2.0	-	--Only one font.
90	U	-	3.2	-	--(Deprecated element in HTML 4.0.)
91	UL	-	2.0	CH	--
		type=disk/circle/square	3.2	-	
		compact	2.0	-	
92	VAR	-	2.0	-	--Only one font.
93	XMP	-	2.0	-	--Only one font. --(Obsolete element in HTML 4.0.)

APPENDIX H

COMPACT HTML DTD

The following code listings (Code Listing H-1) consist of compact HTML *Document Type Definitions* (DTDs).

CODE LISTING H-1 Compact HTML Document Type Definitions[1]

```
<!-- Compact HTML Document Type Definition -->

<!--

Date: Tuesday November 25th 2000

Author: Tomihisa Kamada <tomy@access.co.jp>

-->

<!ENTITY % HTML.Version

"-//W3C//DTD Compact HTML 1.0 Draft//EN"

>

<!--=========== Deprecated Features Switch ===============-->

<!ENTITY % HTML.Deprecated "INCLUDE">

<!--=========== Imported Names ============================-->

<!ENTITY % Content-Type "CDATA">

<!ENTITY % HTTP-Method "GET | POST">

<!ENTITY % URL "CDATA">

<!-- Parameter Entities -->

<!ENTITY % heading "H1|H2|H3|H4|H5|H6">

<!ENTITY % list "UL | OL |  DIR | MENU">

<!ENTITY % preformatted "PRE">
```

continues

CODE LISTING H-1 Compact HTML Document Type Definitions *(continued)*

```
<!--========= Character mnemonic entities ===============-->
<!ENTITY % ISOlat1 PUBLIC
"ISO 8879-1986//ENTITIES Added Latin 1//EN//HTML">
%ISOlat1;
<!--========= Entities for special symbols ==============-->
<!ENTITY amp    CDATA "&"  -- ampersand    -->
<!ENTITY gt     CDATA ">"  -- greater than -->
<!ENTITY lt     CDATA "<"  -- less than    -->
<!--========= Text Markup =================================-->
<!ENTITY % phrase "DFN">
<!ENTITY % special "A | IMG | BR ">
<!ENTITY % form "INPUT | SELECT | TEXTAREA">
<!ENTITY % text "#PCDATA | %phrase | %special | %form">
<!ELEMENT (%phrase) - - (%text)*>
<!ELEMENT BR    - O EMPTY>
<!ATTLIST BR
clear (left|all|right|none) none
>
<!--========= HTML content models ========++===========-->
<!ENTITY % block
"P | %list | %preformatted | DL | DIV | CENTER |
BLOCKQUOTE | FORM | HR ">
<!ENTITY % flow "(%text | %block)*">
<!--========= Document Body ==============================-->
<!ENTITY % body.content "(%heading | %text | %block )*">
<!ELEMENT BODY O O %body.content>
<!ELEMENT DIV - - %body.content>
```

```
<!ATTLIST DIV
align   (left|center|right) #IMPLIED
>
<!ELEMENT center - - %body.content>
<!--=========== The Anchor Element ======================-->
<!ELEMENT A - - (%text)* -(A)>
<!ATTLIST A
name    CDATA   #IMPLIED
href    %URL    #IMPLIED
>
<!--=========== Images ==================================-->
<!ENTITY % Length "CDATA">
<!ENTITY % Pixels "NUMBER">
<!ENTITY % IAlign "(top|middle|bottom|left|right)">
<!ELEMENT IMG    - O EMPTY>
<!ATTLIST IMG
src     %URL     #REQUIRED
align   %IAlign  #IMPLIED
width   %Pixels  #IMPLIED
height  %Pixels  #IMPLIED
hspace  %Pixels  #IMPLIED
vspace  %Pixels  #IMPLIED
alt     CDATA    #IMPLIED
border  %Pixels  #IMPLIED
>
<!--=========== Horizontal Rule =======================-->
<!ELEMENT HR    - O EMPTY>
<!ATTLIST HR
```

continues

CODE LISTING H-1 Compact HTML Document Type Definitions *(continued)*

```
align (left|right|center) #IMPLIED
size  %Pixels #IMPLIED
width %Length #IMPLIED
noshade (noshade) #IMPLIED
>
<!--========= Paragraphs ===================================-->
<!ELEMENT P     - O (%text)*>
<!ATTLIST P
align  (left|center|right) #IMPLIED
>
<!--========= Headings =====================================-->
<!ELEMENT ( %heading )  - -  (%text;)*>
<!ATTLIST ( %heading )
align  (left|center|right) #IMPLIED
>
<!--========= Preformatted Text ===========================-->
<!ENTITY % pre.exclusion "IMG">
<!ELEMENT PRE - - (%text)* -(%pre.exclusion)>
<!--========= Block-like Quotes ===========================-->
<!ELEMENT BLOCKQUOTE - - %body.content>
<!--========= Lists ========================================-->
<!ELEMENT DL - -  (DT|DD)+>
<!ELEMENT DT - O  (%text)*>
<!ELEMENT DD - O  %flow;>
<!ELEMENT (OL|UL) - -  (LI)+>
<!ELEMENT (DIR|MENU) - -  (LI)+ -(%block)>
<!ELEMENT LI - O %flow>
```

```
<!--========= Forms ====================================-->
<!ELEMENT FORM - - %body.content -(FORM)>
<!ATTLIST FORM
action %URL #IMPLIED
method (%HTTP-Method) GET
enctype %Content-Type; "application/x-www-form-urlencoded"
>
<!ENTITY % InputType
"(TEXT | PASSWORD | CHECKBOX | RADIO | HIDDEN
| IMAGE | SUBMIT | RESET )">
<!ELEMENT INPUT - O EMPTY>
<!ATTLIST INPUT
type %InputType TEXT
name  CDATA   #IMPLIED
value CDATA   #IMPLIED
checked (checked) #IMPLIED
size CDATA    #IMPLIED
maxlength NUMBER #IMPLIED
src   %URL    #IMPLIED
align %IAlign #IMPLIED
>
<!ELEMENT SELECT - - (OPTION+)>
<!ATTLIST SELECT
name CDATA #REQUIRED
size NUMBER #IMPLIED
multiple (multiple) #IMPLIED
>
<!ELEMENT OPTION - O (#PCDATA)*>
<!ATTLIST OPTION
```

continues

CODE LISTING H-1 Compact HTML Document Type Definitions *(continued)*

```
selected (selected) #IMPLIED

value  CDATA  #IMPLIED

>

<!ELEMENT TEXTAREA - - (#PCDATA)*>

<!ATTLIST TEXTAREA

name CDATA #REQUIRED

rows NUMBER #REQUIRED

cols NUMBER #REQUIRED

>

<!--========= Document Head =============================-->

<!ENTITY % head.content "TITLE & ISINDEX? & BASE?">

<!ELEMENT HEAD O O  (%head.content)>

<!ELEMENT TITLE - -  (#PCDATA)*>

<!ELEMENT BASE - O EMPTY>

<!ATTLIST BASE

href %URL  #REQUIRED

>

<!ELEMENT META - O EMPTY>

<!ATTLIST META

http-equiv  NAME    #IMPLIED

>

<!--========= Document Structure ========================-->

<!ENTITY % version.attr "VERSION CDATA #FIXED
'%HTML.Version;'">

<!ENTITY % html.content "HEAD, BODY">

<!ELEMENT HTML O O  (%html.content)>

<!ATTLIST HTML
```

```
%version.attr;

>

<!--========= End of DTD =================================-->
```

END NOTES

[1]*Compact HTML for Small Information Appliances*, AnywhereYouGo.com, 3000 Waterview Parkway, B2E14, Richardson, TC 75080, 2001.

APPENDIX I

GLOSSARY

1G First generation of mobile wireless, which utilizes analog air-interface technology.

1XRTT Technology for introducing packet data to IS-95 CDMA to migrate the standard toward 3G.

2G Second generation of mobile wireless, which utilizes various digital protocols, including GSM, CDMA, TDMA, iDEN, and PDC.

2.5G Interim step to 3G involving overlay of higher-capacity data transmission capability to existing 2G digital wireless networks.

3G Third generation wireless employing wideband frequency carriers and a CDMA air interface.

A & B PCS Blocks The first two PCS licenses that were auctioned by the FCC in March 1995. Each contains 30 MHz of spectrum in the 1,900 MHz band and is based on MTA geographic partitions. Mostly large, existing telecommunications companies purchased the licenses.

A-Band Cellular Original mobile wereless frequency band allocated to an entity other than the local telephone company.

Access line A telephone line reaching from the telephone company's central office to a point on private premises. Usually equates to one customer line.

Air-Interface The standard operating system of a wireless network, which is used to communicate to and from the base stations and the handset; technologies include AMPS, TDMA, CDMA, and GSM.

ALS *Alternate Line Service*. Allows you to have two different phone numbers and lines on one subscription. This way you can have separate numbers for business and personal calls. This service is handset and network dependent.

AM *Amplitude Modulation*. CW modulation using amplitude variation in proportion to the amplitude of the modulating signal; usually taken as DSB-LC for commercial broadcast transmissions and DSB-SC for multiplexed systems.

AMPS *Advanced Mobile Phone Service*. Another name for the North American analog cellular phone system.

AMPS roaming The ability to allow or a feature allowing cellular service in a service area or area other than the one in which your contract service originates. Usually known simply as roaming.

Analog In telecommunications, a shortened version of the term analog transmission, which is a way of sending signals—voice, video, or data—in which the transmitted signal is analogous to, or like, the original signal. In other words, if you spoke into a microphone wired to an oscilloscope, and fed your voice coming through a phone line into another oscilloscope, the two signals would look essentially the same. The only difference is that the signal from the phone line would be of a higher frequency.

Antenna A metallic rod that typically extends from a wireless phone and the cell site from which the electrical signal emanates.

AOC *Advice of Charge*. Allows you to monitor the cost of calls made on your mobile phone by displaying the details of the last call and total calls. This service is handset and network dependent.

ARPU *Average Revenue Per User*. A common metric in the wireless industry, this represents the average monthly bill for a customer. It is calculated as total service revenue in a period divided by the average number of subscribers during the period.

Attenuation The loss of signal energy due to absorption, reflection, or diffusion during transmission.

Bandwidth In telecommunications, the width of a communications channel, which is directly proportional to the amount or volume of communications that can travel over that channel. In this case, it is also known colloquially as the size of the pipe.

BASIC authentication A function that verifies the identity of a person accessing a server by demanding input of ID and password.

B-Band cellular Original mobile wireless frequency band allocated to the local telephone company.

Bluetooth Bluetooth is a low power radio technology being developed with the objective of replacing the wires currently used to connect electronic devices such as personal computers, printers, and a wide variety of handheld devices, such as palm top computers and mobile phones.

BPS *Bits-Per-Second*. The standard measure of transmission speed related to the amount of digital computer data capable of being transmitted down a particular channel.

bps/Kbps/Mbps In telecommunications, *bits per second*, *kilobits-per-second* (thousands of bits-per-second) and *megabits-per-second* (millions of bits-per-second); bps is a measure of the transmission speed of data communications in WANs and LANs.

Brand identity The recognizability or memorability of a brand or brand name, as well as the personality, value, quality, dependability, and all other attributes associated with that brand or brand name.

Broadband Also called wideband. Transmission facility whose bandwidth is greater than that available on voice-grade facilities.

BSC *Base Station Controller*. The part of the wireless system's infrastructure that controls one or multiple cell sites' radio signals, thus reducing the load on the switch. Performs radio signal management functions for base transceiver stations, managing functions such as frequency assignment and handoff.

BTA *Basic Training Area*. One of 493 geographic regions in the United States that are used as license areas in the PCS frequency for license blocks C, D, E, and F. Several BTAs make up one MTA.

BTS *Base Transceiver Station*. The name for the antenna and radio equipment necessary to provide wireless service in an area. Also called a base station or cell site.

Bundle Group of pre-paid units.

Bundling The practice of combining several services or products in a single item, sale, or account to make it more attractive to customers.

Call barring Allows you to set up your phone to bar certain incoming or outgoing calls. This service is network and subscription dependent.

Call divert Allows you to divert incoming calls to another mobile phone or answering service. This service is handset and network dependent.

Caller ID A service that allows the recipient of a call to see the source of the call (both the number and the name) on an LCD or LED before answering.

Calling party pays A system in which the person making a call to a wireless phone pays for the call (whether the call is placed from a wireless or a wireline phone). In most countries (the United States is the major exception), calling party pays is the standard.

Call transfer Allows you to transfer a caller to another phone. This service is handset and network dependent.

Carrier A company that provides communications circuits.

C-Block The third PCS license that was auctioned by the FCC in May, 1996. Each contains 30 MHz of spectrum in the 1,900 MHz band and is based on BTA geographic partitions. The licenses were reserved for small businesses and entrepreneurs, and the auction winners were given favorable financing terms.

CDMA *Code Division Multiple Access.*

CDMA2000 Third-generation evolution of cdmaOne that includes several steps, the first of which is 1XRTT.

CDPD *Cellular Digital Packet Data.* An overlay network for transmitting wireless data over first- and second-generation networks.

Cell The basic geographical unit of a cellular phone system. It derives its name from the honeycomb pattern of cell site installations.

Cell broadcast Allows you to receive general messages broadcast in a particular cell such as district information, weather, or financial updates. This service is network and handset dependent.

Cell site Short for cellular site. A location unit through which radio links are established between the wireless system and a wireless unit. A cell site consists of a transmitter/receiver, antenna tower, transmission radios, and radio controllers. A cell site is operated by a *Wireless Service Provider* (WSP).

Cell splitting The process of creating more coverage and capacity in a wireless system by having more than one cell site cover a particular amount of geography. Each cell site covers a smaller area, with lower power MHz, and thus offers the ability to reuse frequencies more times in a larger geographic coverage area, such as a city or MTA.

Cellular The colloquialism for *Cellular Mobile Telephone Service* (or *System*), also known as CMTS.

CF *Call Forwarding*. An intelligent network service that allows you to redirect incoming calls to an alternative destination.

Channel An electrical, electromagnetic, or optical path for communication between two points.

C-HTML *Compact HTML*. A variant of HTML that is supported by i-mode technology.

cHTML *Compact HTML*. Subset of HTML 2.0, HTML 3.2, and HTML 4.0 specifications designed for limited hardware information appliances.

cHTML gateway A cHTML gateway is a software entity within the mobile network. It connects to the Internet or an intranet in order to allow content and applications to be sent to WAP or cHTML enabled devices. Such gateways are expected to be able to handle a number of different markup languages.

Churn Occurs when subscribers cancel service or switch providers. The churn rate is usually expressed by the percentage of total customers who cancel service during a month.

CLEC *Competitive Local Exchange Carrier*. A *local exchange carrier* (LEC) is any telephone company that offers service in a specific area. Now that the industry has been deregulated, several companies may offer service in a single area. New ones entering a market are Competitive Local Exchange Carriers. The original telephone company at the time of deregulation is known as the *Incumbent Local Exchange Carrier* (see also "ILEC").

CLI *Calling Line Identity*. The number of the person calling you is displayed on your phone so that you can choose whether or not to answer the call. Some sophisticated systems, pre-programmed with the identity of certain numbers, will display the name of the caller instead. Some phones have the option to withhold your number when you make a call; this is known as *Calling Line Identity Restriction* (CLIR).

Cloning An illegal theft of cellular phone service. To clone a phone, someone using electronic scanning equipment captures the codes sent out by a cellular telephone and programs them into another cellular telephone. Calls made on that clone phone are then billed to the original phone's account.

CMRS *Commercial Mobile Radio Service*. An FCC designation for any carrier or licensee whose wireless net-work is connected to the public switched telephone network and/or is operated for profit.

Codec Short for coder and decoder, also *Coder-Filter-Decoder* (COFIDEC). Translates audio to digital signals and digital back to audio signals. This is usually accomplished with an *analog-to-digital* (A/D) and *digital-to-analog* (D/A) converter, 25KHz.

Connectivity Enables the sending of information over the Internet.

Control channel The channels that, instead of supporting voice communications, cellular base stations use to continuously broadcast information to cellular phones in the area.

Core business(es) A company's central and main types of business, the one(s) that a company is most proficient at and relies most heavily upon.

Coverage area The geographic area covered by one carrier. If you travel outside a carrier's area, you cannot receive service from that carrier.

Crosstalk Interference in a wireless communications system that stems from other conversations in nearby cells using the same channel.

CTIA *Cellular Telecommunications and Internet Association*. A trade group representing cellular, PCS enhanced specialized mobile radio carriers.

CW *Call Waiting*. Tells you when you are on the phone that another call is waiting.

D-AMPS *Digital AMPS*. Used by Ericsson Inc. to describe IS-136 time division multiple access technology.

D & E PCS Blocks The fourth and fifth PCS licenses that were auctioned by the FCC in January, 1997. Each contains MHz of spectrum in the 1,900 MHz band and is based on BTA geographical partitions.

DAP *Data Access Point*. A Bluetooth term.

Data Any transmittable information other than analog voice.

Data Communications The movement of encoded information by means of electrical transmission systems.

DCS-1800 A variation of the European GSM technology, which forms the basis of what is called a *Personal Communication Network* (PCN). DCS-1800 operates at a higher frequency (1.8 GHz) and can accommodate more users than original GSM networks.

DECT *Digital European Cordless Telephone*. Standard-based on a micro-cellular radio system that provides low-power cordless access between subscriber and base station up to a few hundred meters.

Designated entity The term used in the FCC auctions to describe companies that were eligible to participate in the Central Block auctions, which were reserved for small businesses and entrepreneurs.

Digital The use of binary code to represent information. For analog signals like voice, music, or effects, the sound is sampled many times per second and assigned a number for each sample. Digital technology reproduces sound exactly, and can even filter out background and electronic noise.

Digital Cellular Using a digital protocol over cellular frequencies.

Dropped call When the sender of a cellular call is moving, they often leave one cell (geographical area) and enter another. As a caller moves into another cell, another service provider picks up the call and continues it. This is called a handoff, and usually happens so seamlessly that the caller doesn't even know it's happened. A dropped call is when the service provider in the cell a caller is entering does not pick up the handoff and the call is cut off.

DSL *Digital Subscriber Loop*. A new technology that uses the original telephone company copper wires leading to and from customers' homes to deliver high-speed data services, approximately 300 times faster than normal analog transmissions.

DTMF *Dual Tone Multi-Frequency*. Pulse tone signals transmitted by your phone to communicate with tone activated phone systems such as answering machines and phone banking.

Dual band Dual band refers to the capability of GSM network infrastructure and handsets to operate across two frequency bands. Dual band technology enables a network operator with spectrum at both 900 MHz and 1,800 MHz to support the seamless use of dual band handsets across both frequencies. Dual band networks can provide major benefits in terms of capacity enhancement, greater roaming possibilities for users, and revenue optimization through the introduction of new services.

Dual mode A phone that operates on both analog and digital networks is said to be dual mode. It can be cellular/analog or digital PCS/analog.

EDGE *Enhanced Data for GSM Evolution*. EDGE, which is currently being standardized within the *European Telecommunications Standards Institute* (ETSI), represents the final evolution of data communications within the GSM standard. EDGE uses a new modulation schema to enable data throughput speeds of up to 384 Kbps using existing GSM infrastructure.

EFR *Enhanced Full Rate*. A new speech rate added in some networks to improve audio quality so that it is comparable to that of a land line.

Encryption Method of scrambling voice and data used in digital networks to ensure that conversations and messages cannot be intercepted.

Entrepreneurial A behavior of a person or a business that is independent, free-thinking, individual, and growth-oriented in nature.

EPOC Symbian's technology for wireless information devices including applications, connectivity, and software development. Symbian.com.

ESMR *Enhanced Specialized Mobile Radio*. Wireless service characterized by combining cellular-like feeds with one-to-one and one-to-many dispatch capabilities.

ESN Each cellular phone is assigned a unique *Electronic Serial Number* (ESN), which is automatically transmitted to the cellular base station every time a call is placed. The MTSO validates the ESN with each call. Cloned cellular phones transmit a stolen ESN and charges are made to the real cellular phone account.

FCC *Federal Communications Commission*. A Federal organization established by the Communications Act of 1934 that regulates interstate radio, television, wire, satellite, wireless, and cable communications.

Fiber optic Using fine, transparent lines for the transmission of data, digitally encoded into pulses of light. In terms of telephone conversations, a 1/2" copper cable can transmit about 25 conversations analog, whereas a 1/2" fiber optic line can typically transmit 193,536 conversations digitally.

Fixed dialling numbers Only allows calls to numbers that have been predefined. This service is handset and network dependent.

Foliage attenuation Reductions in signal strength or quality due to signal absorptions by trees or foliage obstructions in the signal's line-sight path. For example, 800 MHz systems are seldom deployed in forested areas—pine needles (nearly the same length as 800 MHz antennas) can negatively affect signal reception in that band.

F PCS Block The final PCS license that was auctioned by the FCC in January, 1997. Each contains 10 MHz of spectrum in the 1,900 MHz band and is based on BTA geographic partitions. The licenses were reserved for small businesses and entrepreneurs, and the auction winners were given favorable financing terms from the FCC.

Frequency reuse The ability of specific channels assigned to a single cell to be used again in another cell, when there is enough distance between the two cells to prevent cochannel interference from affecting service quality. The technique enables a cellular system to increase capacity with a limited number of channels.

Full duplex Both parties on the phone can talk at the same time.

GGSN *Gateway GPRS Support Node*. Interface between the GPRS wireless data network and other networks such as the Internet, X-25, or private networks.

GHz *Gigahertz*. One billion hertz, or cycles per second. Used to measure bandwidth.

GPRS *General Packet Radio Service*. GPRS, which has been standardized as part of the GSM Phase 2+ development, represents the first implementation of packet switching within GSM, which is essentially a circuit switched technology. Rather than sending a continuous stream of data over a permanent connection, packet switching only utilizes the network when there is data to be sent. Using GPRS will enable users to send and receive data at speeds of up to 115 Kbps.

GPS *Global Positioning System*. Refers to satellite-based radio positioning systems that provide 24 hour three-dimensional position, velocity, and time information to suitably equipped users anywhere on or near the surface of the Earth (and sometimes off the Earth). GPS technology is used in a wide range of applications, including maritime, environmental, navigational, tracking, and monitoring.

Gross Adds *Gross Additional Subscribers*. The total number of customers that signed on for service during a period.

GSM *Global System for Mobile communication*. The European standard for digital cellular telephony defined by ETSI. Its implementation is not confined to Europe, but covers a large area of the world and is almost regarded as the de facto world standard.

Half rate Used in GSM to double the amount of channels available.

Handoff The process occurring when a wireless network automatically switches a mobile call to an adjacent cell site.

Handset subsidy Frequently, a wireless company will sell a phone (handset) below cost, with the hope of making up the loss later on customer usage fees. The amount of loss per handset is called the handset subsidy.

HDML *Handheld Device Markup Language*.

HSCSD *High Speed Circuit Switched Data*. GSM is currently a circuit switched technology and HSCSD is the final evolution of circuit switched data within the GSM environment. HSCSD will enable the transmission of data over a GSM link at speeds of up to 57.6 Kbps. This is achieved by concatenating (adding together, consecutive GSM timeslots, each of which is capable of supporting 14.4 Kbps. Up to four GSM timeslots are needed for the transmission of HSCSD.

HTML *Hypertext Markup Language* [HTML4].

HTML 1.0 I-Mode compatible HTML, which supports all i-Mode terminals.

HTML 2.0 I-Mode compatible HTML, which supports only the NTT DoCoMo 502i series terminals.

HTTP *Hypertext Transfer Protocol* [RFC2068].

Hz *Hertz*. A unit of measurement equal to one cycle per second, or one radio wave passing one point in one second of time. Named in honor of Heinrich Hertz, the discoverer of the theory of radio waves.

IDEN *Integrated Digital Enhanced Network.* Digital wireless access technology developed by Motorola to enable cellular-like capabilities on an ESMR-based network. It combines two-way radio, telephone, text messaging, and data transmission into one network.

ILEC *Incumbent Local Exchange Carrier.* A *local exchange carrier* (LEC) is any telephone company that offers service in a specific area. Now that the industry has been deregulated, several companies may offer service in a single area. The original telephone company at the time of deregulation is the Incumbent Local Exchange Carrier. The new ones entering the market are known as *Competitive Local Exchange Carriers* (CLEC).

IMEI *International Mobile Equipment Identifier.* Unique identification number put on every handset manufactured throughout the world.

i-Mode Packet based information service for mobile phones from NTT DoCoMo (Japan). First to provide Web browsing from mobile phones.

IMT 2000 *International Mobile Telecommunications-2000.* Also known as the third-generation mobile systems.

Information Superhighway The original name (and concept) for the Internet.

Internet An information network connecting all modem-equipped computers via telecommunications lines.

IP *Internet Protocol.* The protocol used to break up information into packets, route packets through the network, and reassemble the packets at the destination.

IPv6 *Internet Protocol Version 6.* This is the next generation protocol replacing IPv4. IPv6 will solve the problem of the growing shortage of IPv4 addresses and will be an improvement in areas such as routing and network autoconfiguration.

IrDA A standard for wireless, infrared transmission systems between computers and mobile phones. With IrDA ports, a laptop or PDA can exchange data with a desktop computer or use a printer without a cable connection. IrDA requires line-of-sight transmission like a TV remote control.

ISDN *Integrated Services Digital Network*. The first level of dedicated digital communications lines. Available for homes and offices, ISDN is a big pipe (possesses significantly more bandwidth) that allows video conferencing, many ultra-sophisticated telecommunications features, and extremely fast data communications.

ITN *Independent Telephone Network*. The telephone companies not affiliated with any of the Bell telephone companies.

ITU *International Telecommunications Union*. The United Nations body that recommends 3G standards.

Java A high-level programming language developed by Sun Microsystems to run on most operating systems. Using small Java programs (called applets) a Web site can introduce interactivity and animation.

KHz *Kilohertz*. One thousand hertz, or cycles per second.

Kiosk A free-standing retail location, often in the traffic pattern at malls.

Land line Voice, video, and data transmission technology that relies on wires. See also POTS and Wireline.

LCD *Liquid Crystal Display*. Such as the screen found on an i-Mode cellular phone used for reading e-mail and accessing the Internet.

Local loop The physical wires that run from the customers' telephone, PBX, or key telephone system to the telephone company's central office.

Local service footprint Also known as local service area. The geographical area that a customer may call without incurring toll charges.

Local switch The switch to which a customer's computer telephony system is directly connected, providing better-quality connections.

MHz *Megahertz*. One million hertz, or cycles per second.

Microbrowser A Web browser specialized for a phone or PDA. It is optimized to run in the low-memory and small-screen environment of a handheld device.

Microcell A cell having a very small coverage area, which could be as small as one floor of an office building, one part of an airline terminal, or one corner of a busy intersection. These cells are typically used when coverage and/or capacity is strained and the use of a normal sized cell would cause interference or would be impractical to install. These cells transmit with extremely low power outputs.

MML *Man Machine Language*.

MSA *Metropolitan Statistical Area*. One of 306 geographic regions in the United States that are used as license areas in the cellular frequency band. MSAs are primarily urban areas.

MTA *Major Trading Area*. One of 51 geographic regions in the United States that are used as license areas in the PCS frequency for license blocks A and B. One MTA can be broken up into several BTAs.

MTSO *Mobile Telephone Switching Office*. Central office housing computer processors that handle mobile call switching and interconnection to the wire-based public network. The electronic middleman between call sites and the public switched telephone network, processing traffic back and forth.

Multiparty calls A conference call where it is possible to connect several users to the same line.

Multipath propagation Signal distortion resulting when a transmitted radio-frequency signal is reflected from nearby surfaces on its way to a receiver. The ghosting effect on a TV screen illustrates the multipath phenomenon.

Multiplexing The processes by which several phone calls are carried in the same frequency band at the same time. In wireless, major multiplexing methods include TDMA and CDMA.

M-WorldGate Logica's proposed cHTML gateway.

NACN *North American Cellular Network.* An organization of cellular providers that allows cellular calls across the country to be linked for seamless roaming.

Net Adds *Net Additional Subscribers.* The net new subscribers that a company attracts during a period, calculated as gross adds minus disconnects.

Network Any collection and connection of like or similar services, devices, or companies.

Network services Telecommunications services above and beyond telephone services, such as monitoring of remote sites and equipment for corporations, or transmitting machinery malfunction alerts, burglar alarms, etc.

Non-Wireline carrier Sometimes called the A-Band Cellular carrier, this refers to the original cellular licenses that were granted to companies that were not the local exchange carrier in the region. Because many cellular licenses have been bought and sold since they were first issued in the 1980s, the name is not necessarily accurate any longer.

North American GSM Alliance GSM is the *Global System for Mobile Communications*, a standard adopted by more than 85 countries around the world. The North American GSM Alliance is the group of companies in North America who have adopted similar standards applicable to North American cellular and digital systems.

NTT DoCoMo Japanese cellular provider and chief developer of i-Mode.

OHG *Operators Harmonization Group.* A worldwide organization of operators and manufacturers dedicated to achieving a uniform standard for third-generation wireless systems.

Online Available or accessible through a computer.

Packet switching service A communications system whereby data is divided and transmitted in packets of set sizes. Its special feature is that communication between terminals with differing speeds and formats is possible since transmission/reception is performed after data has first been stored at the exchange.

PBX/Wireless PBX *Private Brand Exchange*. An internal, privately-owned (by an individual or an individual company) telephone or telecommunications system. Large companies have had these for years. Wireless PBX is either cellular (for analog) or digital.

PC card Previously known as PCMCIA card. A credit card sized device that connects laptops or organizers to your phone and acts as a modem to allow data and fax transmission.

PCIA *Personal Communication Industry Association*. A trade group representing PCS, SMR, private radio, and other wireless users and carriers.

PCS/Digital PCS *Personal Communications Services*. Digital wireless telephone service that is lower powered and has higher frequency than regular cellular phones. PCS phones can be less expensive and offer higher sound quality. The downside is that the PCS network is not as far-reaching as the cellular network yet. Digital PCS is a redundancy, as all PCS are digital, but the phrase is used in marketing to differentiate PCS from Cellular.

PDA *Personal Digital Assistant*.

PDC *Personal Digital Cellular*. An international cellular system that uses both full, and half-rate speech code (5.6 Kbps) and allows high-speed transmission at 9.6 Kbps to ensure efficient spectrum utilization.

Penetration The total number of subscribers for a carrier divided by the number of POP that it serves, expressed as a percentage.

Penetration rate The percentage of a market that has been acquired; as in cellular, the percentage of the potential cellular customers that are subscribers to cellular or a specific cellular provider.

PHS *Personal Handyphone System*.

PIN *Personal Identification Number*. A code used by GSM phones to establish authorization for gaining access to all services and information.

POP *Persons of Population*. Wireless industry term for the number of potential subscribers within the licensed area of a cellular PCS system.

Post-pay customers Subscribers that pay for the usage they incur after they incur it. In this case, traditional cellular customers who use their cellular service, then pay for that amount of usage.

POTS *Plain Old Telephone Service*. The lowest-performance voice, video, and data transmission technology. See also wireline and land line.

Prepaid wireless Many carriers offer prepaid wireless services, whereby the customer pays in advance for a certain amount of usage (minutes). Once that time is used up, the customer needs to refill the account before being able to place more calls.

Pre-pay customers Subscribers that pay for the usage they incur before they incur it. In this case, cellular customers pay a fee which is programmed into their account, then use their cellular telephones for the amount of service they have paid for. Very similar to a pre-paid telephone calling card, but for a particular cellular or PCS account.

Promotional messages Advertisements of special pricing, features, or bundles.

Promotional pricing Reduced pricing (for regular service or service bundled with special features) meant to entice new customers to sign up or old customers to come back.

PSTN *Public Switched Telephone Network*. Traditional land line network that mobile wireless systems connect to in order to complete calls.

PUK *PIN Unlocking Key*. If your PIN code is entered incorrectly three times, the SIM card becomes locked. A PUK is then required to unlock it and these can only be obtained from operators.

Repeater Devices that receive a radio signal, amplify it, and retransmit it in a new direction. Used in wireless networks to extend the range of base station signals, thereby expanding coverage more economically than by building additional base stations. Repeaters typically are used for buildings, tunnels, or difficult terrain.

RF *Radio Frequency*. The spectrum of electromagnetic energy between audio and light—500 KHz to 300 GHz.

Roaming The ability of a mobile phone to make and receive calls outside of it's own network. This means you can use networks other than your own when travelling abroad.

Round-up calls/billing When calls are billed by the minute, any call that uses a portion of a minute is rounded up and billed for the whole minute. For example, if you make a call that lasts three minutes and two seconds, you are billed for a four-minute call. Compare to Hello-to-Goodbye billing and True Per-Second BillingSM.

RSA *Rural Statistical Area*. One of 428 geographic regions in the United States that are used as license areas in the cellular frequency band. RSAs are primarily rural areas.

Scratch pad memory The ability to enter numbers into the keypad while having a conversation. This allows you to enter a phone number for use after you have finished talking.

SD card *Secure Digital Card*. The SD Memory Card is a memory device about the size of a postage stamp. The SD card is a joint collaboration between SanDisk, Toshiba, and Panasonic. The SD card offers a combination of high storage capacity (32MB & 64MB in 2000, with the promise of up to 128MB and 256MB in the not-too-distant future), fast data transfer, flexibility, and security. Ideal for downloading files like music, photos, or news to mobile phones, PCs, and other electronic devices.

SGSN *Serving GPRS Support Node*. GRPS receiver of packet transmission from mobile devices at network base stations.

Short messaging service A service that allows short, alpha-numerical messages to be sent to cellular phone display panels.

SIM card *Subscriber Identity Module card*. This is an electronic identity card for digital cellular systems such as GSM. Based on smart card technology, the SIM card enables any handset to take on the identity of a particular user for the purpose of making and receiving calls and for billing. It provides the user with the necessary authentication to access the network and store the GSM encryption algorithms that ensure speech security.

SIM toolkit *Subscriber Identity Module application toolkit*. The SIM toolkit extends the role of the SIM card, making it a key interface between the mobile handset and the network. Using the SIM toolkit, the SIM card can be programmed to carry out new functions and services. These include the ability to manipulate the menu structure of the mobile terminal to provide new, tailored options and download new ringtones, provide local information, etc. This feature is both network and handset dependent.

Smart antenna An antenna system whose technology enables it to focus its beam on a desired signal to reduce interference. A wireless network would employ smart antennas at its base stations in an effort to reduce the number of dropped calls, improve call quality, and improve channel capacity.

SMR *Specialized Mobile Radio*. A technology that provides dispatch service (walkie-talkie-type service used by taxis, delivery trucks, etc.). SMR providers in the United States operate in the 800 MHz and 900 MHz frequency bands and are primarily small, local companies.

SMS *Short Message Service*. The sending and receiving of short text messages of up to 160 characters to mobile phones via the network operator's message center.

Social strata Socio-demographic groups.

Soft handoff Procedure in which two base stations—one in the cell site where the phone is located and the other in the cell site to which the conversation is being passed—both hold onto the call until the handoff is completed. The first cell site does not cut off the conversation until it receives information that the second is maintaining the call.

Spectrum The name given to a range of frequencies used by a wireless carrier to provide service, measured in hertz.

Spread spectrum Radio transmission technology that spreads information over greater bandwidth than necessary for interference tolerance.

Standby time The total amount of time a fully charged battery can power a phone without it being used for a call.

Store-within-a-store Like a kiosk, but by agreement and contract, situated within another company's retail establishment.

Subscribers Contract customers of a specific service.

Superconducting When electricity passes through metal, as signals of any kind, it has to plow through the wire's atoms, which slows it down and diminishes its strength (resistance). The term Superconducting applies to metals, components, or equipment that, through special technology and/or temperature, allow electricity (signals) to pass through with virtually no loss in speed or strength.

Switch A mechanical, electrical, or electronic device that opens or closes circuits, completes or breaks an electrical path, or selects paths or circuits.

Symbian Symbian is a joint venture between Psion, Panasonic, Ericsson, Nokia, and Motorola to develop the EPOC operating system for wireless information devices. A 32-bit operating environment, EPOC comprises a suite of applications, customizable user interfaces, connectivity options, and a range of development tools.

T9™-text input *Predictive Text*. A method of typing text into your keypad using just one key press per letter rather than scrolling through the letters.

Talk time The total amount of time a fully charged battery can be used to power the phone while being continuously used for a call.

TD-CDMA A 3G proposal combining elements of TDMA and CDMA.

TDMA *Time Division Multiple Access*. The system of allocating a number of calls to a cellular channel by dividing them up using time slots.

Telecommunications Communicating by telephone, telegraph, or radio technology. See also telephony.

Telephony Originally meaning voice (analog) communication by telephone (land line), this term has come to encompass virtually all of telecommunications because virtually all of telecommunications can be done over or while connected to a telephone line.

Third-Generation A new standard that promises to offer increased capacity and high-speed data applications up to 2 Mb. It also will integrate pico-, micro-, and macro-cellular technology, and could allow for global roaming.

Traffic This marketing term refers to people actually walking by (a store-within-a-store) or into a location (a store).

Tranceiver Equipment component responsible for the broadcast and reception of radio signals with network or subscriber equipment.

Tri-Mode A phone that operates on analog in the cellular band and digital in both the cellular and PCS bands.

True Per-Second BillingSM Aerial Communications' service mark for Hello-to-Goodbye billing, in which a customer is billed for the exact time between when a recipient of a call picks up the receiver and when one of the parties hangs up. Compare to Round-up billing.

UMTS *Universal Mobile Telecommunications System*. The European member of the IMT2000 family of third generation cellular mobile standards. The goal of UMTS is to support networks that offer true global roaming and can support a wide range of voice, data, and multimedia services. Data rates offered by UMTS are: vehicular—144 Kbps, pedestrian 384 Kbps, and in-building 2 Mbps. Commercial UMTS networks are expected in 2001 in Japan and 2002 in Europe.

Universal Service Fund Administered by the *National Exchange Carrier Association* (NECA) and overseen by the Federal Communications Commission, this is a cost-allocation mechanism designed to keep local exchange rates reasonable, especially in high-rate (read low-profit) areas like rural America.

User agent An HTTP header (HTTP_USER_AGENT) that can identify the model of an accessing i-Mode terminal.

URL *Uniform Resource Locator* [RFC2396].

VoIP *Voice over Internet Protocol*. A technology for transmitting ordinary telephone calls over the Internet using packet-linked routes. Also called IP telephony.

W3C *World Wide Web Consortium*.

WAP *Wireless Application Protocol*.

W-CDMA *Wideband Code Domain Multiple Access*. The radio technology behind UMTS mobile phones. The higher data rates available will allow picture messaging, web browsing and video calls.

Web site Data stored in a server hooked to a modem that you can access by indicating that program's address in an Internet browser program on your modem-equipped computer. A Website can be anything from family vacation photos to entire sales catalogs and purchasing systems.

Wireless Without wires, or any telecommunications that use broadcast (radio) technology versus copper wires (land lines). Most typically, cellular, or digital communications.

Wireline Technology that relies upon wires (i.e., traditional telephony). See also POTS and Land line.

Wireline carrier Sometimes called the B-Band Cellular carrier, this refers to the original cellular licenses that were granted to the LEC in the region. Because many cellular licenses have been bought and sold since they were first issued in the 1980s, the name is not necessarily accurate any longer.

WML *Wireless Markup Language.*

WSP *Wireless Service Provider.* Any company that provides cellular or PCS service.

WWW *World Wide Web.*

INDEX

C

D

E

F

G

H

I

J

K–L

M

N

O

P

Q–R

S

T

U

V

W–Z

RECEIVED
MAR 18 2002